Deepen Your Mind

身為一名網路工程師，你有沒有在工作中遇到過類似這樣的 3 個需求？

需求 1：某大型企業的生產網路裡有 5000 台思科交換機，最近公司更換了 TACACS 伺服器，將思科的 ACS 取代成了 ISE，因為 ISE 的 IP 位址和 ACS 的不同，你需要在這 5000 台交換機上為 ISE 做相關的 AAA 設定，並移除原有的 ACS 的 AAA 設定。

需求 2：公司的網路安全部門提醒你這 5000 台思科交換機現有的 IOS（Internetwork Operating System，網際網路作業系統）版本有很多安全性漏洞，需要儘快升級它們的 IOS 版本。

需求 3：公司聘請的技術稽核人員在隨機抽樣檢查了一些現有的交換機設定後，發現還有很多交換機的設定需要做安全強化和最佳化，例如部分交換機在 Line VTY 的設定下依然允許透過 Telnet 協定遠端存取，部分交換機沒有按要求設定 DHCP Snooping 和 Spanning Tree Portfast 等，你的上司讓你在最短的時間內從這 5000 台交換機裡找出哪些需要做安全強化和最佳化，列出它們的 Hostname 和 IP 位址，以及詳細說明它們各自需要強化和最佳化哪些設定。

時間回到 2013 年，剛剛考取 CCIE 的我第一次聽說軟體定義網路（Software Defined Network，SDN）。當時我在國外某技術討論區讀到一篇關於 SDN 的發文，作者把 SDN 寫得神乎其神，中心思想就是：完全依靠傳統網路工程師手動設定和手動校正，效率不佳的傳統網路運行維護遲早會迎來「壽終正寢」的一天，取而代之的就是能帶來「革命性改變」的 SDN。讀完該文後，作為傳統網路工程師的我感覺到一股強烈的危機感，當時自己花了很長時間去學習和研究 Mininet 這個以 OpenFlow 為主、輕量級的 SDN 模擬器。在研究了一段時間的 Mininet 後，除了對 OpenFlow 有一些了解，我並沒有感覺到 SDN（OpenFlow）為自己從事的網路運行維護工作帶來什麼實質性的幫助和改變。公司裡思科路由器和交換機的設定依然需要一台一台登入去改，IOS 作業系統的升級依然需要一台一台地手動去做；當對裝置的設定、硬體類型和序號做稽核檢查時，依然需要一台一台登入裝置去執行各種 show 指令，對輸出結果用肉眼去篩選和檢查。這些完全依靠網路工程師人工的傳統運行維護工作方式不僅效率不佳，而且容易出現人為失誤，造成不必要的麻煩。自己一度對 SDN 嗤之以鼻，認為該技術的應用場景太過侷限，傳統企業網連線層的運行維護依然離不開網路工程師的手動管理。

2016 年，在新加坡工作將近 7 年後，我有幸受聘於沙烏地阿拉伯阿布都拉國王科技大學（KAUST），從東到西跨越整個亞洲來到這個對很多人來説既熟悉又陌生的國度，在這裡擔任進階網路工程師一職。面對 KAUST 近萬台思科裝置的龐大網路，看著網路運行維護組的同事依然日復一日地用手動的方式一台一台透過 SSH 登入裝置去完成開篇提到的 3 個需求及其他重複、單調、費時的日常工作，我認為需要做些改變來打破傳統，提升工作效率，因為你無法想像每隔半年就需要透過人工手動給超過 5000 台思科交換機升級 IOS 是一項多麼龐大並且費時的工程。以此為契機，在工作之餘，我花了近半年的時間從零開始自學了 Python，並在 KAUST 的生產網路裡進行了實作，寫了大大小小幾十個網路運行維護自動化的指令稿，最後成功透過 Python 語言實現了全面的網路運行維護自動化，相當大地加強了 KAUST 網路運行維護組的工作效率和準確率，讓同事有更多的時間和精力學習其他與電腦網路相關的專業技能，幫助他們從傳統網路工程師向 NetDevOps 工程師轉型。

眾所皆知，Python 這門程式語言的應用場景十分廣泛，人工智慧、資料分析、爬蟲、Web 開發、遊戲製作等領域都能看到 Python 的身影，隨著近幾年 Python 的大熱，與上述 Python 應用場景相關的書籍、視訊、網站等教學資源隨處可見，而 Python 的基礎入門教學更是多如牛毛。遺憾的是，在如此豐富的 Python 教學資源中，為網路工程師量身打造的電腦網路運行維護方面的教學書籍卻相當匱乏。市面上有部分以 NetDevOps 為主題、說明網路運行維護自動化技術的書籍，這種書籍會走馬觀花地把 Linux、Bash、XML/JSON/YANG、NETCONFIG、Jinja、StackStorm 等技術都講一遍，雖然其中有關於 Python 的篇幅，但是內容有限，說明不夠深入，沒有系統地以網路運行維護工作中的實戰程式深入淺出地說明 Python 在大型網路運行維護中的實際應用。很多想學 Python 的網路工程師在讀完這種書籍後依然一頭霧水，不知道怎樣將 Python 運用到工作中。還有一種系統說明 Python 的基礎入門教材，則把 Python 講得過於詳細，其中有很多基礎知識在網路運行維護中很少用到，甚至根本用不到，導致讀者學習週期過長，學習效果也不夠理想。另外，這種教材的作者都是專業程式設計師出身，術業有專攻，他們對網路技術、網路運行維護的了解不如專業網路工程師透徹，自然也就不可能量身打造地寫一本適合網路工程師學習的 Python 教學。

有鑑於此，作為網路工程師出身、已經從事網路運行維護 10 年的我決定結合自己在學習 Python 和在工作中使用 Python 時累積的心得和經驗，寫一本為網路工程師量身打造的 Python 教學，以幫助所有希望轉型或正準備轉型的傳統網路工程師，讓大家在學習網路運行維護自動化技術的道路上少走一些彎路。

另外，Python 核心團隊已經宣佈從 2020 年 1 月 1 日起不再對 Python 2 提供維護和社區支援，由於 Python 3 不相容 Python 2，因此本書所有內容將以截稿前最新為基礎的 Python 3.8.2 做示範。鑑於部分讀者具有一定 Python 2 的基礎，本書也會對 Python 2 和 Python 3 有明顯差異的技術點做說明和示範。

❏ 本書內容簡介

本書共 6 章,分別介紹以下內容。

第 1 章 Python 的安裝和使用

工欲善其事,必先利其器。本章將詳細介紹 Python 在 Windows 和 Linux 作業系統上的安裝和使用方法。

第 2 章 Python 基本語法

為網路工程師量身打造的 Python 程式設計基礎知識的詳細說明是本書的重點內容,分為基本語法(本章)和進階語法(第 3 章)兩部分。本章主要介紹 Python 的變數、方法與函數、資料類型等基礎內容。

第 3 章 Python 進階語法

承接第 2 章的內容,本章將說明 Python 中的條件(判斷)敘述、循環敘述、文字檔的讀 / 寫、自訂函數、模組、正規表示法及異常處理等網路工程師必須掌握的 Python 進階基礎知識。

第 4 章 Python 網路運行維護實驗(GNS3 模擬器)

第 4 章和第 5 章將分別以實驗和實戰的形式說明 Python 在網路運行維護中的實際應用。本章共分為 4 個實驗,實驗難度循序漸進,所有實驗都將在 GNS3 模擬器上示範,實驗程式難度由淺入深,配合詳細的說明,幫助讀者學習和了解。

第 5 章 Python 網路運行維護實戰(實機)

本章提供 3 個在生產網路裡的裝置上實戰執行的 Python 程式說明和示範。每個 Python 指令稿都將提供詳細的分段說明,並且提供指令稿執行前、指令稿執行中、指令稿執行後的畫面,幫助讀者清晰、直觀地了解 Python 是如何把繁雜、單調、耗時的傳統網路運行維護工作實現自動化的。

第 6 章 Python 協力廠商模組詳解

在第 4、5 章的基礎上，本章舉例介紹更多實用的以 Netmiko 為基礎的 Python 協力廠商模組在網路運行維護中的應用，如 TextFSM、ntc-template、Napalm、pyntc、netdev 等，說明它們如何幫助沒有 API 的老舊網路裝置實現更多網路運行維護自動化的功能，如何透過單執行緒非同步及多執行緒的方式加強 Python 指令稿的執行效率。

❏ 適合讀者群

本書適用於熟練掌握了電腦網路技術基礎知識，並且希望學習以 Python 為代表的網路運行維護自動化技術的網路工程師、網路安全工程師、網路顧問、網路架構師，以及電腦網路專業的在校學生。本書也適合已經具備一定 Python 程式設計基礎，並且對網路運行維護自動化技術有興趣的 Linux/Windows 系統工程師和系統架構師。

❏ 本書特色

本書是為網路工程師量身打造的 Python 學習教學，本身就是網路工程師的作者提煉和精選了一些適合網路工程師學習的 Python 基礎知識來說明，幫助對包含 Python 在內任何程式語言都是零基礎的傳統網路工程師快速學習和上手 Python。本書並不是一本全方位的 Python 教學，畢竟術業有專攻，網路工程師不等於全職軟體開發人員。

另外，本書會列出 Python 各種專業術語對應的英文詞彙，例如字串（String）、程式縮排（Indentation）、異常處理（Exception Handling）等，方便讀者在延伸學習和查詢與 Python 相關的英文資料時能快速適應。

最後，本書所有程式都將在 Linux（CentOS）作業系統上執行，並在思科裝置上進行示範。

首先感謝我的親人，感謝你們對我學業、事業的支援，一步一步陪伴我走到今天。感謝你們長期以來默默地在生活及其他方面對我的照顧和關懷，讓我能夠專心完成此書。感謝剛滿三歲的兒子弈仁，你的笑容化解了我生活中的一切煩惱。

感謝在新加坡 12 年的學習、生活、工作中給予過我無私幫助的同學、朋友、同事們。感謝王淵浩和 Lawrence Lee，感謝你們在大學四年同窗及畢業後的生活中給予我各方面的鼓勵和幫助，讓獨自遠離家鄉出國留學的我並不感到孤獨。感謝 Newmedia Express 的老闆馬來西亞人 Alan Woo 和 Shirley Lee，畢業後的半年時間裡求職四處碰壁的那段日子是我人生中最灰暗的一段回憶，感恩你們在我人生最低谷時給了我一份證明我自己的工作機會，它改變了我的人生軌跡。感謝 Wired-Media 公司的新加坡前輩楊紹鵬（Kenneth Yeo），感謝你當年每晚不辭辛勞地繞路駕車送上晚班的我回家，給了我這個剛剛踏入社會、遠離家鄉在獅城打拼的小職員很多溫暖。感謝來自馬來西亞的摯友盧忠聲，相識十年來，與你在工作和生活中一起努力、相互勉勵、相持而笑的日子是我一生中最珍貴的回憶。感謝新加坡同事 Darry Tan，感謝你在我任職於蘋果公司的那段時間裡對我在工作和技術上無私的指導和幫助，感謝你毫無保留地同我分享了你備考 JNCIE、CCDE 的所有筆記和資料，以前輩的身份同我分享了許多寶貴的人生經驗，感恩自己有幸能遇到像你這樣真正以德服人、充滿正能量的導師和貴人。

感謝我在 KAUST 的主管 Gary Corbett、Khalid Mustafa 及 Kevin Sale，沒有你們在工作中及工作外給予我充分的支援、關心和信任，我將無法從一名傳統網路工程師轉型成為 NetDevOps 工程師，自然也就沒有本書的誕生。

I would like to express my sincere gratitude by dedicating this book to Khalid Mustafa，Kevin Sale, Gary Corbett and whoever works with me at King Abdullah University of Science and Technology, without your selfless and continuous support, guidance and encouragement, this book won't be born.

最後感謝所有致力於在電腦網路這一行默默傳授知識和分享經驗的每一個人，你們改變了整個世界！

目錄

01 Python 的安裝和使用

02 Python 基本語法

03 Python 進階語法

04 Python 網路運行維護實驗（GNS3 模擬器）

05 Python 網路運行維護實戰（實機）

06 Python 協力廠商模組詳解

Python 的安裝和使用

欲善其事，必先利其器，鑑於很多網路工程師讀者都是第一次接觸 Python，本書開篇將配圖詳細介紹 Python 在不同作業系統下的安裝和使用方法。Python 在 Windows、Linux 及 MacOS 下都可以使用，目前最新的 MacOS 本身已經內建了 Python，開啟命令列終端輸入指令 python 即可使用。本章主要介紹 Python 在 Windows 和 Linux（CentOS）下的安裝和使用方法。

Python 的執行模式大致分為兩種：一種是使用解譯器（Interpreter）的互動模式（Interactive Mode），另一種是執行指令稿的指令稿模式（Script Mode）。使用解譯器和指令稿來執行 Python 最大的區別是前者能在你執行一行或一段程式後提供「即時回饋」，讓你看到是否獲得了想要的結果，或告訴你程式是否有誤，而後者則是將整段程式寫入一個副檔名為 .py 的文字檔中「包裝執行」。指令稿模式在實際的網路運行維護工作中很常見，但是從學習的角度來講，一定是能提供「即時回饋」的解譯器更利於初學者，因此本章大部分內容將以解譯器為基礎的互動模式來說明，當然，也有某些程式案例必須用指令稿模式來進行示範。

1.1 安裝 Python

本書所有內容以 Windows 10.0（64 位元）和 CentOS 8 分別作為 Windows 和 Linux 兩大作業系統的示範平台。

1.1.1 在 Windows 下安裝 Python 3.8.2

首先在 Python 官網下載 Windows 版的 Python 3（注意，從 Python 3.5 開始，Python 3 已經不再支援 Windows XP 及更早版本的 Windows）。讀者可根據本身情況選擇 32 位元和 64 位元版本，安裝檔案有 .zip、.exe 和 .web-based 3 種格式可選，這裡推薦選擇 .exe 格式，如下圖所示。

Files

Version	Operating System	Description	MD5 Sum	File Size	GPG
Gzipped source tarball	Source release		f9f3768f757e34b342dbc06b41cbc844	24007411	SIG
XZ compressed source tarball	Source release		e9d6ebc92183a177b8e8a58cad5b8d67	17869888	SIG
macOS 64-bit installer	Mac OS X	for OS X 10.9 and later	f12203128b5c639dc08e5a43a2812cc7	30023420	SIG
Windows help file	Windows		7506675dcbb9a1569b54e600ae66c9fb	8507261	SIG
Windows x86-64 embeddable zip file	Windows	for AMD64/EM64T/x64	1a98565285491c0ea65450e78afe6f8d	8017771	SIG
Windows x86-64 executable installer	Windows	for AMD64/EM64T/x64	b5df1cbb2bc152cd70c3da9151cb510b	27586384	SIG
Windows x86-64 web-based installer	Windows	for AMD64/EM64T/x64	2586cdad1a363d1a8abb5fc102b2d418	1363760	SIG
Windows x86 embeddable zip file	Windows		1b1f0f0c5ee8601f160cfad5b560e3a7	7147713	SIG
Windows x86 executable installer	Windows		6f0ba59c7dbeba7bb0ee21682fe39748	26481424	SIG
Windows x86 web-based installer	Windows		04d97979534f4bd33752c183fc4ce680	1325416	SIG

安裝過程中有一個很重要的步驟，如下圖中的 "Add Python 3.8 to PATH"，這裡預設是沒有選取的，請務必選取，它會自動幫你設定好環境變數，也就是說將來在你開啟命令列執行 Python 指令稿時，你可以在任意磁碟代號和資料夾下直接輸入指令 python xxx.py 來執行指令稿，而無須輸入 Python 執行程式所在的完整路徑來執行指令稿，例如 C:\Python38\python xxx.py。不要小看這一選項提供的自動環境變數設定，它能幫助 Python 初學者節省很多很多時間！

之後選擇 "Customize installation" 進入自訂安裝，如下圖所示。

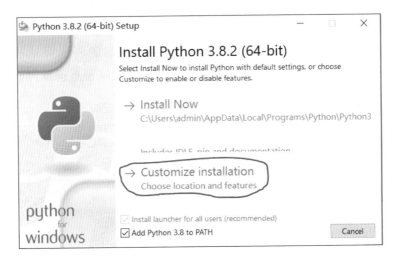

在 Optional Features 的選項中確保 "pip" 和 "tcl/tk and IDLE" 都被選取，關
於它們的作用後面會提到，其他選項使用預設設定即可，然後點擊 "Next"
按鈕，如下圖所示。

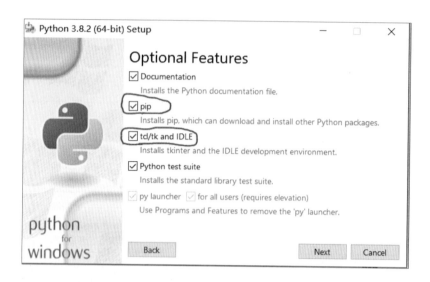

如下圖所示，在 Advanced Options 中，推薦將 "Install for all users" 選取，它會將 Python 的安裝路徑從 C:\Users\admin\AppData\Local\Programs\Python\Python38 換成 C:\Program Files\Python38，方便將來尋找和存取。當然讀者也可以自訂安裝路徑，以及根據本身情況決定是否給所有使用者都安裝 Python 3。

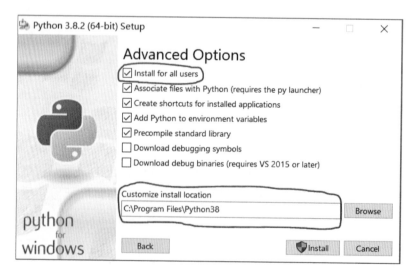

在安裝好 Python 3 後，開啟命令列，輸入 py 或 python，如果可以進入 Python 3.8.2 的解譯器，則說明 Python 3 安裝成功，如下圖所示。

```
C:\Users\admin>
C:\Users\admin>py
Python 3.8.2 (tags/v3.8.2:7b3ab59, Feb 25 2020, 23:03:10) [MSC v.1916 64 bit (AMD64)] on win32
Type "help", "copyright", "credits" or "license" for more information.
>>>
```

註：如果之前你已經安裝過 Python 2，則輸入指令 python 會進入 Python 2，兩個版本之間的使用互不影響。如果只安裝了 Python 3，則只能透過指令 py 來進入 Python 3，如下圖所示。

```
C:\Users\admin>
C:\Users\admin>py
Python 3.8.2 (tags/v3.8.2:7b3ab59, Feb 25 2020, 23:03:10) [MSC v.1916 64 bit (AMD64)] on win32
Type "help", "copyright", "credits" or "license" for more information.
>>> quit()

C:\Users\admin>python
Python 2.7.16 (v2.7.16:413a49145e, Mar  4 2019, 01:37:19) [MSC v.1500 64 bit (AMD64)] on win32
Type "help", "copyright", "credits" or "license" for more information.
>>>
```

1.1.2 在 Linux 下安裝 Python 3.8.2

本書將使用 CentOS 8 作為 Linux 版本的示範平台（在 Windows 上執行的 VMware 虛擬機器），這裡只介紹在 CentOS 命令列終端裡使用 Python 的方法，在 GNOME 桌面環境下使用 Python 的方法不在本書的討論範圍內。因為是實驗環境，所以直接使用 Root 使用者，免去了 sudo 指令，讀者請根據本身情況決定是否使用 sudo。

和 MacOS 一樣，最新的 CentOS 8 已經內建了 Python 2 和 Python 3，輸入 python2 和 python3 兩個指令可以分別進入 Python 2 和 Python 3，如下圖所示。

```
[root@CentOS-Python ~]# python2
Python 2.7.16 (default, Nov 17 2019, 00:07:27)
[GCC 8.3.1 20190507 (Red Hat 8.3.1-4)] on linux2
Type "help", "copyright", "credits" or "license" for more information.
>>> exit()
[root@CentOS-Python ~]#
[root@CentOS-Python ~]# python3
Python 3.6.8 (default, Nov 21 2019, 19:31:34)
[GCC 8.3.1 20190507 (Red Hat 8.3.1-4)] on linux
Type "help", "copyright", "credits" or "license" for more information.
>>> exit()
[root@CentOS-Python ~]#
```

CentOS 8 內建的 Python 3 的版本為 3.6.8，我們需要將它升級到 3.8.2，方法如下。

首先透過下列指令來下載 Python 3.8.2 的安裝套件。

```
wget https://www.python.org/ftp/python/3.8.2/Python-3.8.2.tgz
```

執行該指令後的畫面如下。

```
[root@CentOS-Python ~]# wget https://www.python.org/ftp/python/3.8.2/Python-3.8.2.tgz
--2020-04-27 22:38:45--  https://www.python.org/ftp/python/3.8.2/Python-3.8.2.tgz
Resolving www.python.org (www.python.org)... 151.101.228.223, 2a04:4e42:1a::223
Connecting to www.python.org (www.python.org)|151.101.228.223|:443... connected.
HTTP request sent, awaiting response... 200 OK
Length: 24007411 (23M) [application/octet-stream]
Saving to: 'Python-3.8.2.tgz'

Python-3.8.2.tgz                      100%[===================>]

2020-04-27 22:46:44 (49.0 KB/s) - 'Python-3.8.2.tgz' saved [24007411/24007411]
```

然後輸入下列指令來下載安裝 Python 3.8 所需要的環境相依套件。

```
yum install gcc openssl-devel bzip2-devel libffi-devel
```

執行指令後的畫面如下。

```
[root@CentOS-Python ~]# yum install gcc openssl-devel bzip2-devel libffi-devel
Last metadata expiration check: 0:55:06 ago on Mon 27 Apr 2020 10:25:05 PM EDT.
Package gcc-8.3.1-4.5.el8.x86_64 is already installed.
Package openssl-devel-1:1.1.1c-2.el8_1.1.x86_64 is already installed.
Package bzip2-devel-1.0.6-26.el8.x86_64 is already installed.
Package libffi-devel-3.1-21.el8.x86_64 is already installed.
Dependencies resolved.
Nothing to do.
Complete!
[root@CentOS-Python ~]#
```

接下來用 tar 指令對剛才下載的 Python-3.8.2.tgz 套件解壓縮，解壓縮完成後，目前磁碟代號下會多出一個 Python 3.8.2 的資料夾，用 cd 指令進入該資料夾，如下圖所示。

```
[root@CentOS-Python ~]# tar zxf Python-3.8.2.tgz
[root@CentOS-Python ~]# ls
anaconda-ks.cfg  initial-setup-ks.cfg  Python-3.8.2  Python-3.8.2.tgz
[root@CentOS-Python ~]# cd Python-3.8.2/
[root@CentOS-Python Python-3.8.2]#
```

接著依次輸入下列指令來完成 Python 3.8.2 的安裝。

```
./configure --enable-optimizations
make altinstall
```

執行指令後的畫面如下。

```
[root@CentOS-Python Python-3.8.2]# ./configure --enable-optimizations
checking build system type... x86_64-pc-linux-gnu
checking host system type... x86_64-pc-linux-gnu
checking for python3.8... python3.8
checking for --enable-universalsdk... no
checking for --with-universal-archs... no
checking MACHDEP... "linux"
checking for gcc... gcc
checking whether the C compiler works... yes
checking for C compiler default output file name... a.out
checking for suffix of executables...
checking whether we are cross compiling... no
checking for suffix of object files... o
```

```
[root@CentOS-Python Python-3.8.2]# make altinstall
gcc -pthread -c -Wno-unused-result -Wsign-compare -DNDEBUG -g -fwrapv -O3 -Wall    -std=c99 -Wextra
 -I./Include/internal  -I. -I./Include    -DPy_BUILD_CORE -o Programs/python.o ./Programs/python.c
gcc -pthread -c -Wno-unused-result -Wsign-compare -DNDEBUG -g -fwrapv -O3 -Wall    -std=c99 -Wextra
 -I./Include/internal  -I. -I./Include    -DPy_BUILD_CORE -o Parser/acceler.o Parser/acceler.c
gcc -pthread -c -Wno-unused-result -Wsign-compare -DNDEBUG -g -fwrapv -O3 -Wall    -std=c99 -Wextra
 -I./Include/internal  -I. -I./Include    -DPy_BUILD_CORE -o Parser/grammar1.o Parser/grammar1.c
gcc -pthread -c -Wno-unused-result -Wsign-compare -DNDEBUG -g -fwrapv -O3 -Wall    -std=c99 -Wextra
 -I./Include/internal  -I. -I./Include    -DPy_BUILD_CORE -o Parser/listnode.o Parser/listnode.c
gcc -pthread -c -Wno-unused-result -Wsign-compare -DNDEBUG -g -fwrapv -O3 -Wall    -std=c99 -Wextra
 -I./Include/internal  -I. -I./Include    -DPy_BUILD_CORE -o Parser/node.o Parser/node.c
gcc -pthread -c -Wno-unused-result -Wsign-compare -DNDEBUG -g -fwrapv -O3 -Wall    -std=c99 -Wextra
 -I./Include/internal  -I. -I./Include    -DPy_BUILD_CORE -o Parser/parser.o Parser/parser.c
gcc -pthread -c -Wno-unused-result -Wsign-compare -DNDEBUG -g -fwrapv -O3 -Wall    -std=c99 -Wextra
 -I./Include/internal  -I. -I./Include    -DPy_BUILD_CORE -o Parser/token.o Parser/token.c
gcc -pthread -c -Wno-unused-result -Wsign-compare -DNDEBUG -g -fwrapv -O3 -Wall    -std=c99 -Wextra
```

安裝完畢後，輸入指令 python3.8，如果可以進入 Python 的解譯器，則説明 Python 3.8.2 安裝成功，如下圖所示。

```
[root@CentOS-Python ~]# python3.8
Python 3.8.2 (default, Apr 27 2020, 23:06:10)
[GCC 8.3.1 20190507 (Red Hat 8.3.1-4)] on linux
Type "help", "copyright", "credits" or "license" for more information.
>>> exit()
[root@CentOS-Python ~]#
```

註：安裝 Python 3.8.2 並不會覆蓋 CentOS 內建的 Python 3.6.8，使用指令 python3 仍然可以進入 3.6.8 版本，必須使用指令 python3.8 才能進入 3.8.2 版本，如下圖所示。

```
[root@CentOS-Python ~]# python3
Python 3.6.8 (default, Nov 21 2019, 19:31:34)
[GCC 8.3.1 20190507 (Red Hat 8.3.1-4)] on linux
Type "help", "copyright", "credits" or "license" for more information.
>>> exit()
[root@CentOS-Python ~]# python3.8
Python 3.8.2 (default, Apr 27 2020, 23:06:10)
[GCC 8.3.1 20190507 (Red Hat 8.3.1-4)] on linux
Type "help", "copyright", "credits" or "license" for more information.
>>> exit()
[root@CentOS-Python ~]#
```

1.2 在 Windows 下使用 Python 3.8.2

前面提到 Python 執行模式分為使用解譯器的互動模式和執行指令稿的指令稿模式，下面分別舉例介紹這兩種執行模式在 Windows 中的使用方法。

1.2.1 互動模式

在 Windows 下，有兩種方法進入 Python 解譯器來使用互動模式：一種是透過命令列輸入指令 py 或 python 進入解譯器；另一種是開啟 Python 軟體套件附帶的整合式開發環境（IDE），也就是 IDLE。兩種方法進入的解譯器的介面稍有不同，但是功能完全一樣。

1. 使用命令列進入 Python 解譯器

首先來看第一種方法，開啟 Windows 的命令列（CMD），輸入指令 py 或 python 即可進入 Python 解譯器，如下圖所示。

```
C:\Users\admin>py
Python 3.8.2 (tags/v3.8.2:7b3ab59, Feb 25 2020, 23:03:10) [MSC v.1916 64 bit (AMD64)] on win32
Type "help", "copyright", "credits" or "license" for more information.
>>> print ("hello,world!")
hello,world!
>>>
```

我們在 Python 解譯器中輸入第一段程式 print("hello，world!")，解譯器隨即列印出 "hello，world!" 的內容。這種「即時回饋」的特性是互動模式下特有的，指令稿模式下不具備。

註：在 Python 2 中，print ("hello，world! ") 也可以省去括號寫成 print "hello，world! "，但是在 Python 3 中，print 後面的內容必須加上括號，否則 Python 會顯示出錯，提醒你加上括號，如下圖所示。

```
C:\Users\admin>python
Python 3.8.2 (tags/v3.8.2:7b3ab59, Feb 25 2020, 23:03:10) [MSC v.1916 64 bit (AMD64)] on win32
Type "help", "copyright", "credits" or "license" for more information.
>>> print "hello, world!"
  File "<stdin>", line 1
    print "hello, world!"

SyntaxError: Missing parentheses in call to 'print'. Did you mean print("hello, world!")?
>>>
```

2. 使用 IDLE 進入 Python 解譯器

現在介紹使用 IDLE 進入解譯器的方法。以 Windows 10 為例，點擊左下角的「開始」按鈕後搜尋 "idle" 即可找到 IDLE（Python 3.8 64-bit）這個桌面應用程式。

將 IDLE 開啟後會出現如下圖所示的視窗。再次輸入程式 print ('hello，world!')，可以看到解譯器同樣立即列印出 "hello，world!" 的內容，並且預設支援語法和程式反白。

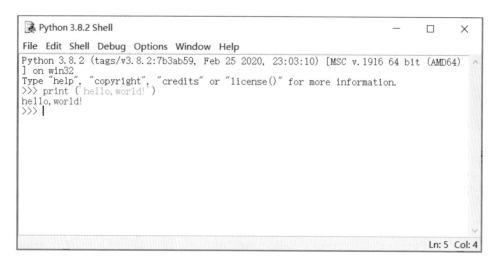

1.2.2 指令稿模式

在 Windows 裡，有兩種方法建立 Python 指令稿，一種是將程式寫進 Windows 記事本裡，另一種是借助協力廠商編輯器。兩種方法分別介紹如下。

1. 使用記事本建立 Python 指令稿

在桌面上新增一個記事本檔案，將程式 print ('hello,world!') 寫入，如下圖所示。

```
print ('hello,world!')
```

然後將其另存為 .py 格式，存在桌面上。這裡需要將「儲存類型」選擇為
「所有檔案」，否則該檔案的類型依然為 .txt。

回到桌面，可以發現第一個 Python 指令稿已經建立成功，如下圖所示。

2. 使用協力廠商編輯器建立 Python 指令稿

支援 Python 的協力廠商編輯器很多，Pycharm、Sublime Text 2/3、Notepad
++、vim（Linux 系統）和 Python 附帶的 IDLE 等都是很優秀也很常用的
編輯器。這裡以 Sublime Text 3 為例簡單介紹使用協力廠商編輯器建立
Python 指令稿的方法。

首先在 Sublime Text 官網下載 Sublime Text 3。Sublime Text 為付費軟體，
但是也可以免費使用，免費版本每使用幾次後會出現一個視窗問你是否願
意購買付費版本，如果你不願意付費，將視窗關閉即可，基本不會影響使
用體驗。

Sublime Text 支援近 50 種程式語言，預設句法（Syntax）是 Plain Text。在
Plain Text 下寫出來的 Python 程式的效果和記事本沒有區別，依然只有黑
白兩色，而且儲存檔案的時候依然需要手動將檔案另存為 .py 格式，如下
圖所示。

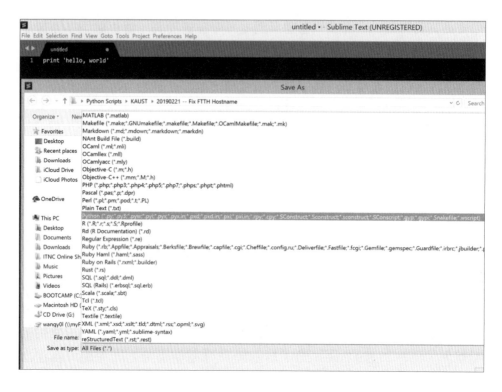

因此，在進入 Sublime Text 後需要做的第一件事是選擇 View → Syntax →
Python 將句法改為 Python，這樣才能獲得對 Python 最好的支援，包含程
式反白、語法提示、程式自動補完、預設將指令稿儲存為 .py 格式等諸多
實用功能，如下圖所示。

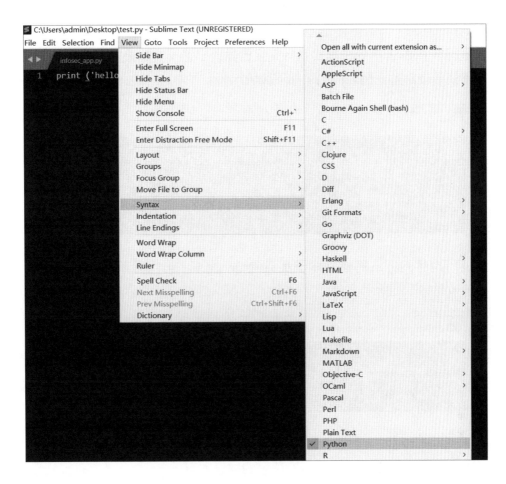

將句法改為 Python 後，程式立刻變為反白，並且儲存程式的時候檔案類型已經自動預設為 .py 格式。

1.2.3 執行 Python 指令稿

在 Windows 系統裡，有四種執行指令稿的方法。

第一種方法是雙擊 .py 檔案，這種方法的缺點是在雙擊執行指令稿後，你會看到一個「閃退」的命令列視窗，視窗閃退速度很快，從出現到消失只有 0.1 ～ 0.2s，肉眼剛剛能看到視窗的輪廓，但是無法看清視窗的內容。
這是因為 Python 指令稿程式執行完後自動退出了，要想讓視窗停留，需要
在程式最後寫上一個 input()，如下圖所示。

然後用同樣的方法將該指令稿另存為 .py 檔案，再次雙擊可執行該指令
稿。效果如下圖所示。

關於 input() 會在 2.4.1 節中詳細解釋，這裡只需要知道可以用它來解決透
過雙擊執行 Python 指令稿時視窗閃退的問題即可。

第二種方法是在命令列裡移動到指令檔所在的資料夾下，輸入 y xxx.py 或
python xxx.py 指令來執行指令稿，如下圖所示。

第三種方法是使用 IDLE 來執行指令稿，實際步驟為：首先使用滑鼠按右
鍵指令檔，選擇 "Edit with IDLE"，進入 IDLE，如下圖所示。

然後在 IDLE 裡點擊 Run → Run Module 來執行指令稿，如下圖所示。

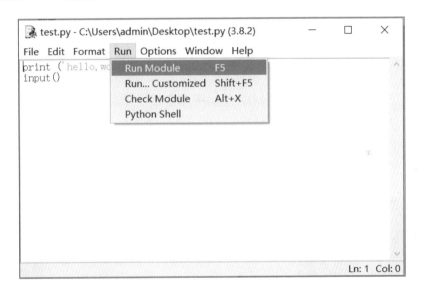

獲得的效果如下圖所示。

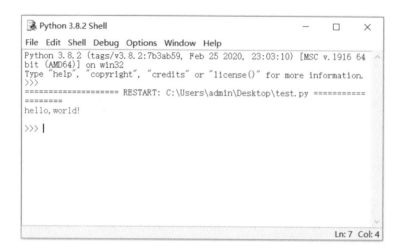

可以發現，在 IDLE 裡即使不使用 input()，執行指令稿時也不會出現視窗閃退的問題，因此，通常建議使用 IDLE 來執行指令稿。

第四種方法是在協力廠商編輯器裡執行指令稿。依然以 Sublime Text 3 為例，方法很簡單，首先進入 Sublime Text 3，如下圖所示，依次選擇 Tools → Build System → Python。

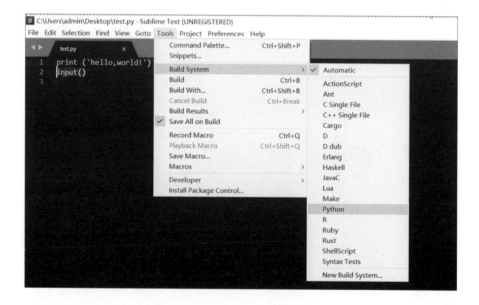

寫好程式並儲存後，開啟 Tools → Build 或使用快速鍵 Ctrl + B 就可以在視窗底部看到執行指令稿的結果了，效果如下圖所示。

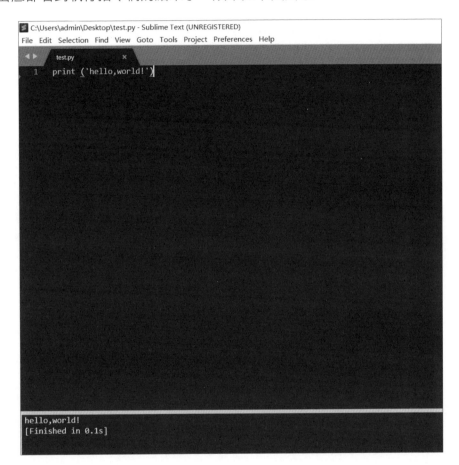

1.3 在 Linux 下使用 Python 3.8.2

前面提到本書只介紹在 CentOS 命令列終端裡使用 Python 的方法，在 GNOME 桌面環境下使用 Python 的方法不在本書的討論範圍內。下面介紹互動模式和指令稿模式在 CentOS 中的使用方法。

1.3.1 互動模式

我們知道，在 CentOS 的命令列終端裡輸入指令 python3.8 即可進入 Python 3.8.2 的解譯器，也就進入了 Python 的互動模式，如下圖所示。

```
[root@CentOS-Python ~]# python3.8
Python 3.8.2 (default, Apr 27 2020, 23:06:10)
[GCC 8.3.1 20190507 (Red Hat 8.3.1-4)] on linux
Type "help", "copyright", "credits" or "license" for more information.
>>>
```

在 Python 解譯器裡輸入第一段程式 print ('hello, world!')，解譯器隨即列印出了 "hello, world!" 的內容。這種「即時回饋」的特性是互動模式下特有的，指令稿模式下不具備，如下圖所示。

```
root@CentOS-Python:~                                      —    □    ×
[root@CentOS-Python ~]# python3.8
Python 3.8.2 (default, Apr 27 2020, 23:06:10)
[GCC 8.3.1 20190507 (Red Hat 8.3.1-4)] on linux
Type "help", "copyright", "credits" or "license" for more information.
>>> print ('hello,world!')
hello,world!
>>>
```

1.3.2 指令稿模式

在 CentOS 的命令列終端裡，我們可以使用文字編輯器來建立指令稿，CentOS 有幾種常見的文字編輯器，如 emacs、nano、vi 等。這裡介紹用 vi 建立 Python 指令稿的方法。

關於 vi 的用法本書將只做簡單介紹。另外，vi 有一個加強版本叫作 vim，兩者的實際區別不在本書的討論範圍內。讀者只需要知道在建立 Python 指令稿時，vim 支援語法高亮度，而 vi 不支援。

vi 不支援語法高亮度,僅顯示 PuTTY 預設字型的顏色。vim 支援語法高亮度,顯示彩色字型。

除此之外,兩者對 Python 的支援並無本質區別,是否喜歡語法高亮度全憑個人喜歡。只需要注意一點:vi 是 CentOS 安裝時附帶的文字編輯器,vim 則需要透過輸入指令 yum install vim 安裝後才能使用。

下面用實例介紹使用 vi 建立 Python 指令稿的方法。

首先輸入指令 vi test.py,建立一個名為 test.py 的 Python 指令稿,如下圖所示。

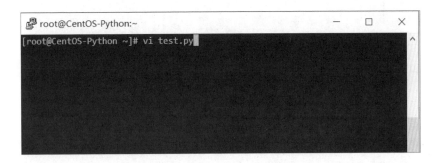

進入 vi 後,按 "i" 鍵進入輸入模式(螢幕左下角會顯示 "- - INSERT - -"),輸入第一段程式 print ('hello, world!'),如下圖所示。

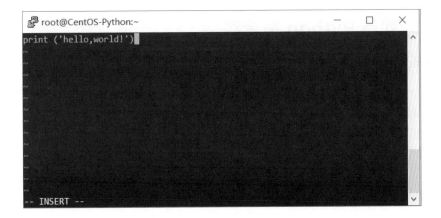

然後按 "ESC" 鍵，在螢幕左下角的 "- - INSERT - -" 消失後，接著輸入 :wq，按確認鍵後即可儲存檔案並退出 vi，如下圖所示。

之後回到命令列終端，輸入 ls 即可看到剛剛建立成功的 Python 指令稿 test.py，如下圖所示。

1.3.3 執行 Python 指令稿

與 Windows 命令列終端一樣，在 CentOS 命令列終端也是透過輸入 python xxx.py 來執行 Python 指令稿的，不同的是，因為本書以 Python 3.8.2 為例，這裡需要把 python 換成 python3.8，寫成 python3.8 xxx.py 來執行指令稿，如下圖所示。

1.3.4 Shebang 符號

在 Linux 和 UNIX 裡，符號 #! 叫作 Shebang，通常可以在 Linux/UNIX 系統指令稿中第一行的開頭看到它。它的作用是指明執行指令檔的解釋程式。寫在 Shebang 後面的解釋程式如果是一個可執行檔，當執行指令稿時，Shebang 會把檔案名稱作為參數傳遞給解釋程式去執行。例如 python3.8 test.py 中的 python3.8 是解釋程式，test.py 是檔案名稱，使用 Shebang 後，可以省去解釋程式，把 python3.8 test.py 寫成 ./test.py 就可以執行 Python 指令稿了。另外，Shebang 指定的解釋程式必須為可執行程式，否則系統會顯示出錯 "Permission denied."。

因此，如果你覺得每次都需要輸入指令 python2、python3 或 python3.8 來執行指令稿比較麻煩，則可以在指令稿的開頭部分使用 Shebang 符號，然後在其後面加上 /usr/bin/env python3 來指定 python3 為解釋程式（同理，如果你想使用 python2 來做解釋程式，則可以寫成 #!/usr/bin/env python2），如下圖所示。

將指令稿儲存並退出後,用 chmod 指令將 test.py 改為可執行,如下圖所示。

然後就可以用 ./test.py 來執行指令稿,省去每次都必須輸入指令 python2、python3 或 python3.8 的麻煩,如下圖所示。

Python 基本語法

本章主要介紹 Python 的基本語法知識，舉例示範的所有程式都將在 CentOS 中執行，如果程式前面帶有 >>> 符號，則為在解譯器下執行的程式，不帶 >>> 符號的程式即為指令稿模式下執行的程式。

2.1 變數

所謂變數（Variable），顧名思義，指在程式執行過程中，值會發生變化的量。與變數相對應的是常數，也就是在程式執行過程中值不會發生變化的量，不同於 C/C++ 等語言，Python 並沒有嚴格定義常數這個概念，在 Python 中約定俗成的方法是使用全大寫字母的命名方式來指定常數，如圓周率 PI=3.1415926。

變數是儲存在記憶體中的值，在建立一個變數後，也就表示在記憶體中預留了一部分空間給它。變數用來指向儲存在記憶體中的物件，每個物件根據本身情況又可以代表不同的資料類型（Data Type）。我們可以透過

變數設定值這個操作將變數指向一個物件，例如下面的 a = 10 即一個最簡單的變數設定值的範例。

```
>>> a = 10
```

在 Python 中，我們使用等號 = 來連接變數名稱和值，進而完成變數設定值的操作。這裡將 10 這個整數（也就是記憶體中的物件）設定值給變數 a，因為 10 本身是「**整數**」（Integer），所以變數 a 此時就代表了「整數」這個資料類型的值。我們可以使用 type() 函數來確認 a 的資料類型，發現變數 a 的資料類型此時為 int，也就是 integer 的縮寫，程式如下。

```
>>> type(a)
<class 'int'>
>>>
```

Python 是一種動態類型語言，和 C、Java 等不同，我們無須手動指明變數的資料類型，根據設定值的不同，Python 可以隨意更改一個變數的資料類型。舉例來說，剛才我們把「整數」這個資料類型的值設定值給變數 a，現在再次設定值一個內容為 test 的**字串**（String）資料類型的值給變數 a，然後用 type() 函數確認，這時 a 的資料類型已經從 int 變為了 str，即字串，程式如下。

```
>>> a = 'test'
>>> type(a)
<class 'str'>
>>>
```

變數名稱可以用大小寫英文字母、底線、數字來表示，但是不能包含標點符號、空格及各種其他特殊符號，如括號、貨幣符號等。

變數名稱可以以字母和底線開頭，**但是不能以數字開頭**，舉例如下。

```
>>> test = 'test'
>>> _a_ = 1
```

```
>>> 123c = 10
  File "<stdin>"，line 1
    123c = 10
       ^
SyntaxError: invalid syntax
>>>
```

這裡 Python 解譯器傳回了 "SyntaxError: invalid syntax" 這個無效語法的錯誤訊息，告訴我們 123c 為無效的變數名稱。這也是使用解譯器來學習 Python 的優勢，無論程式裡出現什麼問題，都能獲得「即時回饋」。

變數名稱區分大小寫，舉例如下。

```
>>> a = 10
>>> print (A)
Traceback (most recent call last):
  File "<stdin>", line 1, in <module>
NameError: name 'A' is not defined
>>> print (a)
10
>>>
```

如果變數名稱中間出現兩個或以上的單字，則只能用底線將它們連接，不可以使用空格將它們隔開，舉例如下。

```
>>> ip_address = '192.168.1.1'
>>> ip address = '192.168.1.1'
  File "<stdin>", line 1
    ip address = '192.168.1.1'
        ^
SyntaxError: invalid syntax
>>>
```

最後，不是所有的英文單字都能用作變數名稱，Python 中有**保留字**（Reserved Word）的概念。保留字通常是 Python 中常用的關鍵字，例如

用作建立函數的 "def"，用作 while 循環和 for 循環的 "while" 和 "for"，等等。可以用下面的方法來查詢目前的 Python 版本中有哪些保留字。

```
>>> import keyword
>>> print (keyword.kwlist)
['False', 'None', 'True', 'and', 'as', 'assert', 'async', 'await', 'break',
'class', 'continue', 'def', 'del', 'elif', 'else', 'except', 'finally',
'for', 'from', 'global', 'if', 'import', 'in', 'is', 'lambda', 'nonlocal',
'not', 'or', 'pass', 'raise', 'return', 'try', 'while', 'with', 'yield']
>>>
```

看不懂上面的程式沒關係，本書後面會講到它們的用法。這裡只需注意輸入 print (keyword.kwlist) 後的傳回值為一個串列，該串列中的元素（串列和元素的概念後文會講到）即目前 Python 版本中的保留字，這些保留字均不能用來做變數名稱，舉例如下。

```
>>> and = 1
  File "<stdin>", line 1
    and = 1
      ^
SyntaxError: invalid syntax
>>> as = 1
  File "<stdin>", line 1
    as = 1
      ^
SyntaxError: invalid syntax
>>> else = 1
  File "<stdin>", line 1
    else = 1
        ^
SyntaxError: invalid syntax
>>>
```

2.2 註釋

在程式設計中，尤其是當撰寫程式內容較多的大型程式時，**註釋**（Comments）可以造成備註的作用，在回顧程式時幫助回憶和了解。除此之外，在團隊合作的時候，在程式中使用註釋也是極其重要的一項要求，因為你寫的程式可能會被他人呼叫、維護，為了讓他人更容易了解你寫的程式的目的和用途，在程式中使用註釋是非常必要的。

在 Python 中，我們使用 # 來做註釋符號。和 # 寫在同一排，並且寫在 # 後面的程式將只做註釋使用，不會被當作程式的一部分，也就不會被執行。大多數情況下我們會在指令稿模式中用到註釋，在互動模式中使用註釋的情況很少，舉例如下。

```
# coding=utf-8
#產生一個整數串列，該串列為整數1~10的平方數的集合
test_list = [i ** 2 for i in range(1,11)]
print (test_list)
```

這段程式也許你看不懂，但是透過註釋可以知道它的作用是「產生一個整數串列，該串列為整數 1 ～ 10 的平方數的集合」，也就是 1，4，9，16，…，100。我們可以執行該指令稿檢視輸出內容。

需要注意的是，如果使用指令稿模式執行 Python，並且程式中出現了中文，則必須在程式的開頭加上 "# coding=utf-8"，因為 Python 預設的編碼格式是 ASCII，如果不修改編碼格式，則 Python 將無法正確顯示中文。

因為寫在 # 後面的程式只做註釋使用，並不會被當作程式的一部分執行，因此有時我們還可以巧用 # 來「遮蓋」我們不想執行的程式。例如可以選擇性地在指令稿的某些 print() 函數前加上 #，不看其輸出內容，想看其輸出內容時再把 # 刪除，而不至於每次都要在指令稿裡反覆刪除、重新定義該段 print() 函數。

2.3 方法和函數

在 Python 中，方法（Method）和函數（Function）大致來説是可以互換的兩個詞，它們之間有一個細微的區別：函數是獨立的功能，無須與物件連結；方法則與物件有關，不需要傳遞資料或參數就可以使用。舉個實例，前面講到的 type() 就是一個函數，程式如下。

```
>>> a = 123
>>> type(a)
<class 'int'>
>>>
>>> type('xyz')
<class 'str'>
>>>
```

方法則需要與一個物件（變數或資料）連結，例如 upper() 是一個方法，它的作用是將字串裡小寫的英文字母轉為大寫的英文字母，程式如下。

```
>>> vendor = 'Cisco'
>>> vendor.upper()
'CISCO'
>>>
```

這裡我們建立了一個名為 vendor 的變數，並將字串內容 "Cisco" 設定值給它，隨後對該變數呼叫 upper() 方法，傳回值即所有字母都變為大寫的 "CISCO"。

在 Python 中，每種資料類型都有自己預設附帶的函數、方法和變數，要檢視某一資料類型本身具有的函數、方法和變數，可以使用 dir() 函數，這裡以字串和整數為例，程式如下。

```
>>> dir(str)
['__add__'、'__class__'、'__contains__'、'__delattr__'、'__doc__'、
```

```
'__eq__'、'__format__'、'__ge__'、'__getattribute__'、'__getitem__'、
'__getnewargs__'、'__getslice__'、'__gt__'、'__hash__'、'__init__'、
'__le__'、'__len__'、'__lt__'、'__mod__'、'__mul__'、'__ne__'、'__new__'、
'__reduce__'、'__reduce_ex__'、'__repr__'、'__rmod__'、'__rmul__'、
'__setattr__'、'__sizeof__'、'__str__'、'__subclasshook__'、'_formatter_
field_name_split'、'_formatter_parser'、'capitalize'、'center'、'count'、
'decode'、'encode'、'endswith'、'expandtabs'、'find'、'format'、'index'、
'isalnum'、'isalpha'、'isdigit'、'islower'、'isspace'、'istitle'、
'isupper'、'join'、'ljust'、'lower'、'lstrip'、'partition'、'replace'、
'rfind'、'rindex'、'rjust'、'rpartition'、'rsplit'、'rstrip'、'split'、
'splitlines'、'startswith'、'strip'、'swapcase'、'title'、'translate'、
'upper'、'zfill']

>>> dir(int)
['__abs__'、'__add__'、'__and__'、'__class__'、'__cmp__'、'__coerce__'、
'__delattr__'、'__div__'、'__divmod__'、'__doc__'、'__float__'、
'__floordiv__'、'__format__'、'__getattribute__'、'__getnewargs__'、
'__hash__'、'__hex__'、'__index__'、'__init__'、'__int__'、'__invert__'、
'__long__'、'__lshift__'、'__mod__'、'__mul__'、'__neg__'、'__new__'、
'__nonzero__'、'__oct__'、'__or__'、'__pos__'、'__pow__'、'__radd__'、
'__rand__'、'__rdiv__'、'__rdivmod__'、'__reduce__'、'__reduce_ex__'、
'__repr__'、'__rfloordiv__'、'__rlshift__'、'__rmod__'、'__rmul__'、
'__ror__'、'__rpow__'、'__rrshift__'、'__rshift__'、'__rsub__'、
'__rtruediv__'、'__rxor__'、'__setattr__'、'__sizeof__'、'__str__'、
'__sub__'、'__subclasshook__'、'__truediv__'、'__trunc__'、'__xor__'、
'bit_length'、'conjugate'、'denominator'、'imag'、'numerator'、'real']
>>>
```

以上即使用 dir() 函數列出的字串和整數所附帶的函數、方法和變數。注意，其中前後都帶單底線或雙底線的變數不會在本書中介紹，例如 "_formatter_parser" 和 "__contains__"，初學 Python 的網路工程師只需要知道它們在 Python 中分別表示私有變數與內建變數，學有餘力的網路工程師可以自行閱讀其他 Python 書籍深入學習，其他不帶底線的函數與方法並不是每一個都會在網路運行維護中用到，筆者將在下一章中選取說明。

2.4 資料類型

前面講到，我們可以使用變數來指定不同的資料類型，對網路工程師來說，常用的資料類型有字串（String）、整數（Integer）、串列（List）、字典（Dictionary）、浮點數（Float）、布林（Boolean）。另外，不是很常用但需要了解的資料類型包含集合（Set）、元組（Tuple）及空值（None）。下面一一舉例說明。

2.4.1 字串

字串即文字，可以用單引號''、雙引號" "和三引號""" ""表示，下面分別介紹三者的區別和用法。

1. 單引號和雙引號

當表示內容較短的字串時，單引號和雙引號比較常用且兩者用法相同，例如 'cisco' 和 "juniper"，需要注意的是**單引號和雙引號不可以混用**。

```
>>> vendor1 = 'Cisco'
>>> vendor2 = "Juniper"
>>> vendor3 = 'Arista"
  File "<stdin>", line 1
    vendor3 = 'Arista"
                     ^
SyntaxError: EOL while scanning string literal
>>> vendor3 = 'Arista'
```

這裡建立了 3 個變數：vendor1、vendor2 和 vendor3，分別將字串 "Cisco"、"Juniper" 和 "Arista" 設定值給它們，因為字串 Arista 混用了單引號和雙引號，導致解譯器顯示出錯，重新給 vendor3 設定值並且只使用單引號後解決了這個問題。這時我們可以用 print 敘述（Statements）將 3 個變數的內容列印出來，如下所示。

```
>>> print (vendor1)
Cisco
>>> print (vendor2)
Juniper
>>> print (vendor3)
Arista
>>>
```

除了使用 print() 函數，我們還可以在解譯器裡直接輸入變數名稱來取得它的值，這是編輯器互動模式下特有的，指令稿模式做不到，舉例如下。

```
>>> vendor1
'Cisco'
>>> vendor2
'Juniper'
>>> vendor3
'Arista'
>>>
```

需要指出的是，如果變數中存在分行符號 \n，則 print 會執行換行動作，但如果在解譯器裡直接輸入變數名稱，則解譯器會把分行符號 \n 當作字串內容的一部分一起傳回，舉例如下。

```
>>> banner = "\n\n  Warning: Access restricted to Authorised users only. \n\n"
>>> print banner


  Warning: Access restricted to Authorised users only.

>>> banner
'\n\n  Warning: Access restricted to Authorised users only. \n\n'
>>>
```

看出區別了嗎？

在 Python 中，我們還可以透過加號＋來連接（Concatenate）字串，舉例如
下。

```
>>> ip = '192.168.1.100'
>>> statement = '交換機的IP位址為'
>>>
>>> print (statement + ip)
交換機的IP位址為192.168.1.100
```

注意，在使用加號＋將變數連接合併時，如果其中一個變數為字串，那麼
其他所有要與之連接的變數也都必須為字串，否則 Python 會顯示出錯，舉
例如下。

```
>>> statement1='網段192.168.1.0/24下有'
>>> quantity = 60
>>> statement2='名使用者'
>>>
>>> print (statement1 + quantity + statement2)
Traceback (most recent call last):
  File "<stdin>"，line 1，in <module>
TypeError: cannot concatenate 'str' and 'int' objects
>>>
```

這裡 statement1 和 statement2 兩個變數都為字串，但是 quantity 這個變數為
整數，因此 print statement1 + quantity + statement2 會顯示出錯 "TypeError:
cannot concatenate 'str' and 'int' objects"，提示不能將字串和整數連接合併。
解決辦法是使用 str() 函數將 quantity 從整數轉化為字串，程式如下。

```
>>> print (statement1 + str(quantity) + statement2)
網段192.168.1.0/24下有60名使用者
>>>
```

2. 三引號

三引號形式的字串通常用來表示內容較長的文字，它最大的好處是如果遇到需要換行的文字，文字內容裡將不再需要分行符號 \n。例如路由器和交換機中用來警告非授權使用者非法存取裝置後果的 MOTD（Message of The Day）之類的旗標（Banner）設定，這種文字內容通常比較長且需要換行，這時用三引號來表示該文字內容是最好的選擇，舉例如下。

```
>>> motd = ''' -----------------------------------------------------------
...
... Warning: You are connected to the Cisco systems，incorporated network.
... Unauthorized access and use of this network will be vigorously prosecuted.
...
... -----------------------------------------------------------'''
>>>
>>> print (motd)
-----------------------------------------------------------------------
Warning: You are connected to the Cisco systems，incorporated network.
Unauthorized access and use of this network will be vigorously prosecuted.
-----------------------------------------------------------------------
>>>
```

3. 與字串相關的方法與函數

❑ upper()

前面提到過 upper() 方法，它的作用是將字串裡小寫的英文字母轉為大寫的英文字母。upper() 的傳回值是字串，舉例如下。

```
>>> vendor = 'Cisco'
>>> vendor.upper()
'CISCO'
>>>
```

❏ lower()

顧名思義，與 upper() 相反，lower() 方法的作用是將字串裡大寫的英文字母轉為小寫的英文字母。lower() 的傳回值是字串，舉例如下。

```
>>> vendor = 'CISCO'
>>> vendor.lower()
'cisco'
>>>
```

❏ strip()

strip() 用來在字串的開頭和結尾移除指定的字元（如字母、數字、空格、分行符號 \n、標點符號等）。如果沒有指定任何參數，則預設移除字串開頭和結尾處的所有空格和分行符號 \n。有時在字串的開頭和結尾會夾雜一些空格，如 "192.168.100.1"，要去掉這些多餘的空格，可以使用 strip()。strip() 的傳回值是字串，舉例如下。

```
>>> ip='   192.168.100.1   '
>>> ip.strip()
'192.168.100.1'
>>>
```

有時在字串尾端會有分行符號 \n（例如使用 open() 函數的 readlines() 方法來讀取文字檔裡的內容後所傳回的串列裡的元素，後面會講到），我們也可以使用 strip() 來移除這些分行符號，舉例如下。

```
>>> ip='192.168.100.1\n'
>>> ip.strip()
'192.168.100.1'
>>>
```

❏ count()

count() 用來判斷一個字串裡指定的字母或數字實際有多少個，例如要找

出 "39419591034989320" 這個字串裡有多少個數字 9，就可以用 count()。
count() 的傳回值是整數，舉例如下。

```
>>> '39419591034989320'.count('9')
5
>>>
```

❑ len()

len() 用來判斷字串的長度，例如要回答上面提到的 "39419591034989320"
總共是多少位數，就可以用 len()。len() 的傳回值是整數，舉例如下。

```
>>> a='39419591034989320'
>>> len(a)
17
>>>
```

❑ split() 和 join()

之所以把這兩個方法放在一起講，是因為他們關係比較接近，在字串、串
列的轉換中互成對應的關係，split() 將字串轉換成串列，join() 將串列轉換
成字串。

到目前為止，我們還沒有講到串列（List），這裡簡單說明一下：在 Python
中，串列是一種有序的集合，用中括號 [] 表示，該集合裡的資料又被叫作
元素，例如 [1,3,5,7,9] 就是一個最簡單的串列，其中的整數 1、3、5、7、
9 都屬於該串列的元素。下面我們把該串列設定值給變數 list1，用 type()
來確認該變數的資料類型，可以發現它的資料類型為 list。

```
>>> list1 = [1,3,5,7,9]
>>> type(list1)
<class 'list'>
```

我們可以使用索引來存取和指定串列中的每個元素，索引的順序是從數字
0 開始的。串列索引的用法舉例如下。

```
>>> list1[0]
1
>>> list1[1]
3
>>> list1[2]
5
>>> list1[3]
7
>>> list1[4]
9
>>>
```

講完串列後，為了配合下面的案例，需要講一下 input() 函數。

- **input() 的傳回值是字串**。它的作用是提示使用者輸入資料與 Python 程式互動，例如你想寫一段程式詢問使用者的年齡，讓使用者自己輸入年齡，可以寫一段這樣的指令稿程式。

```
[root@CentOS-Python ~]# cat demo.py

age = input('How old are you? ')
print ('Your age is: ' + age)

[root@CentOS-Python ~]#
```

然後執行該指令稿程式。

```
[root@CentOS-Python ~]# python demo.py

How old are you? 32
Your age is: 32

[root@CentOS-Python ~]#
```

注意這裡的 32 是使用者自己輸入的，雖然它看著像整數，但是它實際的資料類型是字串。

- 在 Python 2 中，上面提到的 Python 3 的 input() 函數實際對應的是
  Python 2 的 raw_input()，而 Python 2 中的 input() 函數只能用來接收數
  字（整數或浮點數），並且相對應的傳回值也為整數或浮點數。實際
  上 Python 3 將 Python 2 中的 input() 和 raw_input() 整合成了 input()（在
  Python 3 中已經沒有 raw_input()，這點需要注意）。

在了解了串列和 input() 函數大致的原理和用法後，再來看網路工程師如
何在網路運行維護中使用 split() 和 join()。舉例來說，在大中型公司裡，
IP 位址的劃分一般是有規律可循的，比如説某公司有一棟 10 層樓的建
築，一樓的 IP 子網為 192.168.1.0/24，二樓的為 192.168.2.0/24，三樓的為
192.168.3.0/24，依此類推。現在你需要做個小程式，讓使用者輸入任意一
個屬於公司內網的 IP 位址，然後讓 Python 告訴使用者這個 IP 位址屬於哪
一層樓。想法如下：

因為該公司內網 IP 位址的第一段都為 192，第二段都為 168，第四段不管
使用者輸入任何 IP 位址都不影響我們對樓層的判斷。換句話説，我們只能
從該 IP 位址的第三段來判斷是哪一層樓，但是我們要怎樣告訴 Python 去
判斷哪一個數字屬於 IP 位址的第三段呢？這時就可以用 split() 將使用者輸
入的 IP 位址（字串）轉化成串列，然後透過串列的索引來指向 IP 位址的
第三段，程式如下。

```
>>> floor1='192.168.1.0'
>>> floor1_list = floor1.split('.')
>>>
>>> print floor1_list
['192', '168', '1', '0']
>>>
>>> floor1_list[2]
'1'
>>>
```

我們先將 192.168.1.0 設定值給 floor1 這個變數，再對該變數呼叫 split() 方法，然後將傳回值指定給另一個變數 floor1_list。注意 split() 括號裡的 "." 表示分隔符號，該分隔符號用來對字串進行切片。因為 IP 位址的寫法都是 4 個數字用 3 個 "." 分開，所以這裡分隔符號用的是 "."。**因為 split() 的傳回值是串列，所以我們 print floor1_list 後可以看到，IP 位址的 4 個數字已經被切片獨立開來，分別成為組成 floor1_list 串列的 4 個元素的其中之一**，之後我們就可以透過 floor1_list[2] 這種索引的方式來查詢該串列的第三個元素，進一步獲得 IP 位址的第三段的數字了，也就是這裡的數字 1。

在知道怎麼透過 split() 來取得 IP 位址的第三段的數字後，回到前面的需求：讓使用者輸入任意一個屬於公司內網的 IP 位址，然後讓 Python 告訴使用者這個 IP 位址屬於哪一層樓。指令稿程式如下。

```
[root@CentOS-Python ~]# cat demo.py

# coding=utf-8
ip = input('請輸入要查詢的IP位址: ')
ip_list = ip.split('.')
print ('該IP位址屬於' + ip_list[2] + '樓.')

[root@CentOS-Python ~]#
```

我們使用 input() 函數提示使用者輸入想要查詢的 IP 位址，然後將獲得的值（字串）設定值給變數 ip。隨後對其呼叫 split() 函數，並將傳回值（串列）指定給另一個變數 ip_list。接著透過 ip_list[2] 做索引，獲得該串列的第三個元素，也就是使用者輸入的 IP 位址的第三段。最後用 print 將查詢的結果傳回告知使用者。

執行以下程式來看效果。

```
[root@CentOS-Python ~]# python demo.py

請輸入要查詢的IP位址: 192.168.3.100
```

```
該IP位址屬於3樓

[root@CentOS-Python ~]#
```

講完 split() 後，再來看 join() 怎麼用。首先來看下面這個串列，它包含了開啟思科交換機通訊埠的幾條最基本的指令。

```
>>> commands = ['configure terminal', 'interface Fa0/1', 'no shutdown']
```

這幾行指令缺少了關鍵的一點：分行符號 \n（也就是確認鍵），這時我們可以使用 join() 將分行符號 \n 加在每行指令的尾端，注意 join() 的傳回值**是字串**。

```
>>> '\n'.join(commands)
'configure terminal\ninterface Fa0/1\no shutdown\n'
>>>
```

再舉個實例，如果我們要把之前的串列 ['192','168','1','0'] 轉換回字串 "192.168.1.0"，可以採用以下做法。

```
>>> '.'.join(['192','168','1','0'])
'192.168.1.0'
>>>
```

如果不加這個 "." 會怎樣？試試看。

```
>>> ''.join(['192','168','1','0'])
'19216810'
>>>
```

❑ startswith()，endswith()，isdigit()，isalpha()

之所以把上述 4 個字串的函數和方法放在一起講，是因為**它們的傳回值都是布林值（Boolean）**。布林值只有兩種：True 和 False，且字首必須大寫，true 和 false 都不是有效的布林值。布林值通常用來判斷條件是否成立，如果成立，則傳回 True；如果不成立，則傳回 False。

首先來看 startswith()，用來判斷字串的內容是否以指定的字串開頭，舉例如下。

```
>>> ip = '172.16.5.12'
>>> ip.startswith('17')
True
>>> ip.startswith('172.')
True
>>> ip.startswith('192')
False
>>>
```

endswith() 與 startswith() 剛好相反，用來判斷字串的內容是否以指定的字串結尾，舉例如下。

```
>>> ip = '192.168.100.11'
>>> ip.endswith('1')
True
>>> ip.endswith('11')
True
>>> ip.endswith('2')
False
>>>
```

字串的內容包羅萬象，可以為空，可以為中文漢字或英文字母，可以為整數或小數，可以為任何標點符號，也可以為上述任意形式的組合。而 isdigit() 就是用來判斷字串的內容是否為整數的，舉例如下。

```
>>> year='2019'
>>> year.isdigit()
True
>>> vendor='F5'
>>> vendor.isdigit()
False
>>> PI='3.1415926'
>>> PI.isdigit()
```

```
False
>>> IP='1.1.1.1'
>>> IP.isdigit()
False
>>>
```

isalpha() 用來判斷字串的內容是否為英文字母，舉例如下。

```
>>> chinese = '中文'
>>> chinese.isalpha()
False
>>> english = 'English'
>>> english.isalpha()
True
>>> family_name = 'Wang'
>>> family_name.isalpha()
True
>>> full_name = 'Parry Wang'
>>> full_name.isalpha()
False
>>> age = '33'
>>> age.isalpha()
False
>>>
```

注意，isalpha() 很嚴格，只要字串的內容出現了哪怕一個非英文字母，isalpha() 就會傳回 False，如 'Parry Wang'（包含了空格）、'Parry_Wang'（包含了底線）等。

2.4.2 整數和浮點數

在 Python 中，有 5 種數值型態（Numeric Type），分別為整數（Integer）、浮點數（Float）、布林類型（Boolean）、長整數（Long）和複數（Complex），對網路工程師來說，掌握前面三種就夠了，後面兩種不是我們需要關心的。

所謂整數，即通常了解的不帶小數點的正數或負數，浮點數則是帶小數點的正數或負數。可以透過 type() 函數來驗證，以下面的 1 為整數，1.0 則為浮點數。

```
>>> type(1)
<class 'int'>
>>> type(1.0)
<class 'float'>
>>>
```

我們可以把 Python 當成一個計算機，使用 +、-、*、/、//、** 等**算術運算子**做加、減、乘、除、求冪等常見的數學運算，舉例如下。

```
>>> 256 + 256
512
>>> 1.2 + 3.5
4.7
>>> 1024 - 1000
24
>>> 16 * 16
256
>>> 100/10
10
>>> 12 // 10
1
>>> 12 % 10
2
>>> 8**2
64
>>> 3**3
27
>>>
```

在 Python 中，可以透過運算子 ** 做冪運算，例如 8 的 2 次方可以表示為 8**2，3 的 3 次方表示為 3**3。

在做除法運算時，可以看到範例中分別使用了 /、// 和 % 3 個運算子，它們的區別如下。

/ 表示正常的除法運算，注意在 Python 2 中，如果碰到整數除整數，結果出現小數部分的時候，例如 12 / 10，Python 2 只會傳回整數部分，即 1，要想得到小數點後面的部分，必須將除數或被除數透過 float() 函數換成浮點數來運算，舉例如下。

```
[root@CentOS-Python ~]# python2
Python 2.7.16 (default, Nov 17 2019, 00:07:27)
[GCC 8.3.1 20190507 (Red Hat 8.3.1-4)] on linux2
Type "help", "copyright", "credits" or "license" for more information.
>>> 12/10
1
>>> 12/float(10)
1.2
>>> float(12)/10
1.2
>>>
```

而在 Python 3 中，12/10 則直接傳回浮點數 1.2。

```
[root@CentOS-Python ~]# python3.8
Python 3.8.2 (default, Apr 27 2020, 23:06:10)
[GCC 8.3.1 20190507 (Red Hat 8.3.1-4)] on linux
Type "help", "copyright", "credits" or "license" for more information.
>>> 12/10
1.2
>>>
```

// 表示向下取整數，求商數。

```
>>> 12//10
1
>>>
```

% 則表示求餘數。

```
>>> 12%10
2
>>>
```

整數也不單單用來做數學運算，透過加號 + 或乘號 * 兩種運算子，**還可以
與字串互動**，適合用來畫分割線，舉例如下。

```
>>> print ('CCIE ' * 8)
CCIE CCIE CCIE CCIE CCIE CCIE CCIE CCIE
>>>
>>> print ('CCIE ' + 'CCIE')
CCIE CCIE
>>>
>>> print ('*' * 50)
**************************************************
>>>
```

在網路運行維護中，有時會遇到需要用計數器做統計的時候，比如說某公
司有 100 台思科 2960 交換機，由於長期缺乏系統性的運行維護管理，交
換機的 IOS 版本並不統一。為了統計其中有多少台交換機的 IOS 版本是
最新的，需要登入所有的交換機，每發現一台 IOS 版本為最新的交換機就
透過計數器加 1，直到結束。由於要完成這個指令稿需要有關 Paramiko、
if、for 循環、正規表示法等進階性的 Python 基礎知識，所以這裡僅示範計
數器的用法。

```
>>> counter = 0
>>> counter = counter + 1
>>> counter
1
>>> counter = counter + 1
>>> counter
2
>>> counter += 1
```

```
>>> counter
3
>>> counter += 1
>>> counter
4
>>>
```

首先我們建立一個變數 counter，將 0 設定值給它，該變數就是我們最初始
的計數器。之後如果每次發現有交換機的 IOS 版本為最新，就在該計數器
上加 1，注意 counter = counter + 1 可以簡寫為 counter + = 1。

2.4.3 串列

串列（List）是一種有序的集合，用中括號 [] 表示，串列中的資料被叫作
元素（Element），每個元素之間都用逗點隔開。串列中的元素的資料類型
可以不固定，舉例如下。

```
>>> list1 = [2020，1.23，'Cisco'，True，None，[1,2,3]]
>>>
>>> type(list1[0])
<class 'int'>
>>> type(list1[1])
<class 'float'>
>>> type(list1[2])
<class 'str'>
>>> type(list1[3])
<class 'bool'>
>>> type(list1[4])
<class 'NoneType'>
>>> type(list1[5])
<class 'list'>
>>>
```

由上可知，我們建立了一個名為 list1 的變數（注意 list 在 Python 中是保留字，並不能被用作變數名稱，所以我們用 list1 作為變數名稱），並將一個含有 6 個元素的串列設定值給它。可以看到這 6 個元素的資料類型都不一樣，我們使用 type() 函數配合串列的索引來驗證每個元素的資料類型，**串列的索引值從 0 開始，對應串列裡的第 1 個元素**。可以發現從第 1 個到第 6 個元素的資料類型分別為整數、浮點數、字串、布林值、空值，以及串列。

註：一個串列本身也可以以元素的形式存在於另一個串列中，舉例來說，上面的串列 list1 的第 6 個元素為串列 [1,2,3]，我們可以透過使用兩次索引的方法來單獨調取串列 [1,2,3] 中的元素，也就是整數 1、2、3。

```
>>> list1 = [2020, 1.23, 'Cisco', True, None, [1,2,3]]
>>> list1[5][0]
1
>>> list1[5][1]
2
>>> list1[5][2]
3
>>>
```

下面介紹與串列相關的方法和函數

❑ range()

range() 函 數 在 Python 2 和 Python 3 中 有 較 大 區 別。 在 Python 2 中，range() 函數用來建立一個整數串列，傳回值為串列，舉例如下。

```
[root@CentOS-Python ~]# python2
Python 2.7.16 (default, Nov 17 2019, 00:07:27)
[GCC 8.3.1 20190507 (Red Hat 8.3.1-4)] on linux2
Type "help", "copyright", "credits" or "license" for more information.
>>>
>>>a = range(10)
```

```
>>>type(a)
<type 'list'>  #Python 2中的range()函數傳回值的類型為串列
>>> print a
[0，1，2，3，4，5，6，7，8，9]
>>> range(1, 15)
[1，2，3，4，5，6，7，8，9，10，11，12，13，14]
>>> range(1, 20, 2)
[1，3，5，7，9，11，13，15，17，19]
>>>
```

- Python 2 中的 range() 建立的整數串列從 0 開始，因此 range(10) 傳回的是一個包含整數 0 ～ 9 的串列，**並不包含 10**。

- 也可以在 range() 中指定起始數和結尾數，傳回的整數串列的最後一個元素為指定的結尾數減 1，如 range(1, 15) 將傳回一個包含整數 1 ～ 14 的串列，14 由結尾數 15 減 1 得來。

- range() 還可以透過指定步進值來得到我們想要的整數，例如你只想選取 1 ～ 19 中所有的單數，那麼就可以使用 range (1, 20, 2) 來實現（這裡的 2 即步進值）。

這種傳回串列的 range() 函數有一個缺點是會佔用記憶體，串列所含元素數量不多時對主機的效能影響不大，但是當使用 range(10000000000000) 來建立諸如這樣極大的串列時所佔用的記憶體就非常「恐怖」了。因此在 Python 3 中，range() 函數的傳回值被改成 range，這是一種可以被反覆運算的物件，這樣改的目的就是節省記憶體。

```
[root@CentOS-Python ~]# python3.8
Python 3.8.2 (default, Apr 27 2020, 23:06:10)
[GCC 8.3.1 20190507 (Red Hat 8.3.1-4)] on linux
Type "help", "copyright", "credits" or "license" for more information.
>>> a = range(10)
>>> type(a)
<class 'range'> #Python 3中的range()函數的傳回值不再是串列，而是range反覆運算值。
```

```
>>> print (a)
range(0, 10)      #不再像Python 2那樣傳回整數串列[0,1,2,3,4,5,6,7,8,9]
>>> range (1,15)
range(1, 15)         #同上
>>> range (1,20,2)
range(1, 20, 2)    #同上
>>>
```

如果要使 Python 3 的 range() 也傳回串列，則需要對其使用 list() 函數，舉例如下。

```
>>> a = list(range(10))
>>> print (a)
[0, 1, 2, 3, 4, 5, 6, 7, 8, 9]
>>> list(range(1,15))
[1, 2, 3, 4, 5, 6, 7, 8, 9, 10, 11, 12, 13, 14]
>>> list(range(1,20,2))
[1, 3, 5, 7, 9, 11, 13, 15, 17, 19]
>>>
```

❏ append()

append() 用來在串列增加元素，舉例如下。

```
>>> interfaces = []
>>> interfaces.append('Gi1/1')
>>> print (interfaces)
['Gi1/1']
>>> interfaces.append('Gi1/2')
>>> print (interfaces)
['Gi1/1'，'Gi1/2']
>>>
```

首先我們建立一個空串列（以 [] 表示），並把它設定值給 interfaces 變數，然後使用 append() 方法將通訊埠 Gi1/1 加入該串列，隨後呼叫 append() 將 Gi1/2 加入該串列，現在串列 interfaces 中就有 Gi1/1 和 Gi1/2 兩個元素了。

❑ len()

串列的 len() 方法和字串的 len() 方法大同小異，前者用來統計串列中有多少個元素，後者用來統計字串內容的長度，其傳回值也依然為整數，舉例如下。

```
>>> len(interfaces)
2
>>>
>>> cisco_switch_models = ['2960','3560','3750','3850','4500','6500',
'7600','9300']
>>> len(cisco_switch_models)
8
>>>
```

❑ count()

與字串一樣，串列也有 count() 方法，串列的 count() 方法用來找出指定的元素在串列中有多少個，傳回值為整數，舉例如下。

```
>>> vendors = ['Cisco','Juniper','HPE','Aruba','Arista','Huawei','Cisco',
'Palo Alto','CheckPoint','Cisco','H3C','Fortinet']
>>> vendors.count('Cisco')
3
>>>
```

❑ insert()

串列是有序的集合，前面講到的 append() 方法是將新的元素增加到串列的最後面，如果我們想自己控制新元素在串列中的位置，則要用 insert() 方法。舉例如下。

```
>>> ospf_configuration = ['router ospf 100\n', 'network 0.0.0.0
255.255.255.255 area 0\n']
>>> ospf_configuration.insert(0, 'configure terminal\n')
>>> print (ospf_configuration)
```

```
['configure terminal\n', 'router ospf 100\n', 'network 0.0.0.0
255.255.255.255 area 0\n']
>>>
```

首先我們建立一個名為 ospf_configuration 的變數，將設定 OSPF 的指令寫在一個串列中設定值給該變數。隨後發現串列漏了 configure terminal 指令，該指令要寫在串列的最前面，這時我們可以用 insert(0, 'configure terminal\n') 將該指令加在串列的最前面（記住串列的索引值是從 0 開始的）。

如果這時我們還想給該 OSPF 路由器設定一個 router-id，如把 router-id 這行指令寫在 router ospf 100 的後面，可以再次使用 insert()，舉例如下。

```
>>> ospf_configuration.insert(2, 'router-id 1.1.1.1\n')
>>> print (ospf_configuration)
['configure terminal\n', 'router ospf 100\n', 'router-id 1.1.1.1\n', 'network
0.0.0.0 255.255.255.255 area 0\n']
>>>
```

❏ pop()

pop() 用來移除串列中的元素，如果不指定索引值，則 pop() 預設將去掉排在串列尾端的元素；如果指定了索引值，則可以精確移除想要移除的元素，舉例如下。

```
>>> cisco_switch_models = ['2960', '3560', '3750', '3850', '4500', '6500',
'7600', '9300']
>>> cisco_switch_models.pop()
'9300'   #排在尾端的9300被去掉
>>> print cisco_switch_models
['2960', '3560', '3750', '3850', '4500', '6500', '7600']
>>> cisco_switch_models.pop(1)
'3560' #使用索引值1將排在第2位的3560去掉
>>> print (cisco_switch_models)
```

```
['2960', '3750', '3850', '4500', '6500', '7600']
>>>
```

❑ index()

看了 pop() 的用法後，你也許會問：在擁有很多元素的串列中，怎麼知道想要移除的元素的索引值是多少呢？這時就需要用 index()，如想從 cisco_switch_models 串列中移除元素 4500，可以按照以下操作。

```
>>> cisco_switch_models = ['2960', '3560', '3750', '3850', '4500', '6500',
'7600', '9300']
>>> cisco_switch_models.index('4500')
4
>>> cisco_switch_models.pop(4)
'4500'
>>> print (cisco_switch_models)
['2960', '3560', '3750', '3850', '6500', '7600', '9300']
>>>
```

先透過 index() 找出 4500 的索引值為 4，然後配合 pop(4) 將它從串列中移除。

2.4.4 字典

在 Python 裡，字典（Dictionary）是許多組無序的鍵值對（Key-Value pair）的集合，用大括號 {} 表示，每一組鍵值對都用逗點隔開，舉例如下。

```
>>> dict = {'Vendor':'Cisco', 'Model':'WS-C3750E-48PD-S', 'Ports':48,
'IOS':'12.2(55)SE12', 'CPU':36.3}
```

這裡我們建立了一個變數名稱為 dict 的字典，該字典有 5 組鍵值對，分別如下。

```
'Vendor':'Cisco'
'Model':'WS-C3750E-48PD-S'
'Ports':48
'IOS':'12.2(55)SE12'
'CPU':36.3
```

- 鍵值對裡的鍵（Key）和值（Value）用冒號 : 隔開，冒號的左邊為鍵，右邊為值。

- 鍵的資料類型可為字串、常數、浮點數或元組，對網路工程師來說，最常用的一定是字串，如 "Vendor"、"Model" 等。

- 值可為任意的資料類型，例如這裡的 "Cisco" 為字串，48 為整數，36.3 為浮點數。

與串列不同，字典是無序的，舉例如下。

```
>>> a = [1, 2, 3, 'a', 'b', 'c']
>>> print (a)
[1,  2, 3, 'a', 'b', 'c']
>>>
>>> dict = {'Vendor':'Cisco',  'Model':'WS-C3750E-48PD-S',  'Ports':48,
'IOS':'12.2(55)SE12',  'CPU':36.3}
>>> print (dict)
{'IOS': '12.2(55)SE12',  'CPU':36.3,  'Model': 'WS-C3750E-48PD-S',  'Vendor':
'Cisco',  'Ports': 48}
>>>
```

這裡我們建立一個內容為 [1，2，3，'a'，'b'，'c'] 的串列 a，將它列印出來後，串列中元素的位置沒有發生任何變化，因為串列是有序的。但是如果我們將剛才的字典 dict 列印出來，你會發現字典裡鍵值對的順序已經徹底被打亂了，沒有規律可循，正因為字典是無序的，我們自然也不能像串列那樣使用索引來尋找字典中某個鍵對應的值。

在字典裡，尋找某個值的格式為 '字典名 [鍵名]'，舉例如下。

```
>>> dict = {'Vendor':'Cisco', 'Model':'WS-C3750E-48PD-S', 'Ports':48,
'IOS':'12.2(55)SE12', 'CPU':36.3}
>>> print (dict['Vendor'])
Cisco
>>> print (dict['CPU'])
36.3
>>> print (dict['Ports'])
48
>>>
```

如果要在字典裡新增加一組鍵值對，則格式為 '字典名 [新鍵名]' = ' 新值 '，
舉例如下。

```
>>> dict['Number of devices']=100
>>> print (dict)
{'Vendor': 'Cisco', 'Number of devices': 100, 'IOS': '12.2(55)SE12',
'CPU': 36.3, 'Model': 'WS-C3750E-48PD-S', 'Ports': 48}
>>>
```

如果要更改字典裡某個已有鍵對應的值，則格式為 '字典名 [鍵名]' = ' 新
值 '，舉例如下。

```
>>> dict['Model'] = 'WS-C2960X-24PS-L'
>>> dict['Ports'] = '24'
>>> print (dict)
{'IOS': '12.2(55)SE12', 'Model': 'WS-C2960X-24PS-L', 'Vendor': 'Cisco',
'Ports': '24', 'CPU': 36.3}
>>>
```

如果要刪除字典裡的某組鍵值對，則格式為 del '字典名 [鍵名]'，舉例如
下。

```
>>> del dict['Number of devices']
>>> print (dict)
```

```
{'Vendor': 'Cisco',  'IOS': '12.2(55)SE12',  'CPU': 36.3,  'Model':
'WS-C3750E-48PD-S',  'Ports': 48}
>>>
```

下面介紹與字典相關的函數和方法。

❏ len()

len() 用來統計字典裡有多少組鍵值對。**len() 的傳回值是整數**，舉例如下。

```
>>> print (dict)
{'Vendor': 'Cisco',  'IOS': '12.2(55)SE12',  'CPU': 36.3,  'Model':
'WS-C3750E-48PD-S',  'Ports': 48}
>>> len(dict)
5
>>>
```

❏ keys()

keys() 用來傳回一個字典裡所有的鍵。**keys() 在 Python 2 中的傳回值為串列；在 Python 3 中的傳回值是可反覆運算的物件**，需要使用 list() 將它轉為串列，了解即可。舉例如下。

```
>>> print (dict)
{'Vendor': 'Cisco',  'IOS': '12.2(55)SE12',  'CPU': 36.3,  'Model':
'WS-C3750E-48PD-S',  'Ports': 48}
>>> print (dict.keys())
['Vendor',  'IOS',  'CPU',  'Model',  'Ports']
>>>
```

❏ values()

values() 用來傳回一個字典裡所有的值。注意，**values() 在 Python 2 中的傳回值為串列，在 Python 3 中的傳回值是可反覆運算的物件**，在有必要的情況下需要使用 list() 將它轉為串列，舉例如下。

```
[root@CentOS-Python ~]# python2
Python 2.7.16 (default, Nov 17 2019, 00:07:27)
[GCC 8.3.1 20190507 (Red Hat 8.3.1-4)] on linux2
Type "help", "copyright", "credits" or "license" for more information.
>>> dict = {'Vendor':'Cisco', 'Model':'WS-C3750E-48PD-S', 'Ports':48,
'IOS':'12.2(55)SE12', 'CPU':36.3}
>>> print (dict.values())
['Cisco', '12.2(55)SE12', 36.3, 'WS-C3750E-48PD-S', 48] #Python 2裡傳回串列
>>>

[root@CentOS-Python ~]# python3.8
Python 3.8.2 (default, Apr 27 2020, 23:06:10)
[GCC 8.3.1 20190507 (Red Hat 8.3.1-4)] on linux
Type "help", "copyright", "credits" or "license" for more information.
>>> dict = {'Vendor':'Cisco', 'Model':'WS-C3750E-48PD-S', 'Ports':48,
'IOS':'12.2(55)SE12', 'CPU':36.3}
>>> print (dict.values())
dict_values(['Cisco', 'WS-C3750E-48PD-S', 48, '12.2(55)SE12', 36.3]) #Python
3裡傳回可反覆運算物件
>>>
```

❑ pop()

前面講到，要刪除字典中某組鍵值對可以用指令 del ' 字典名 [鍵名]'。另
外，我們可以使用 pop() 來達到同樣的目的。與串列的 pop() 不同，字典的
pop() 不能匯入索引值，需要匯入的是鍵名，**而且字典的 pop() 的傳回值不
是串列，而是鍵名對應的值（例如下面的 48），**舉例如下。

```
>>> print (dict)
{'Vendor': 'Cisco', 'IOS': '12.2(55)SE12', 'CPU': 36.3, 'Model':
'WS-C3750E-48PD-S', 'Ports': 48}
>>> dict.pop('Ports')
48
>>> print (dict)
{'Vendor': 'Cisco', 'IOS': '12.2(55)SE12', 'CPU': 36.3, 'Model':
```

```
'WS-C3750E-48PD-S'}
>>>
```

❑ get()

前面講到，我們可以使用 values() 方法獲得一個字典裡所有的值。除此之外，還可以使用 get() 來傳回字典裡實際鍵名對應的值，**get() 的傳回值是所匯入的鍵名對應的值**，舉例如下。

```
>>> print (dict)
{'Vendor': 'Cisco',  'IOS': '12.2(55)SE12',  'CPU': 36.3,  'Model':
'WS-C3750E-48PD-S'}
>>> dict.get('Vendor')
'Cisco'
>>> dict.get('CPU')
36.3
>>>
```

2.4.5　布林類型

布林類型（Boolean）用來判斷條件是否成立，布林值只有兩種：True 和 False，如果條件成立，則傳回 True；如果條件不成立，則傳回 False。**兩種布林值的字首（T 和 F）必須大寫，true 和 false 都不是有效的布林值。**布林類型在判斷敘述中常用，Python 的判斷敘述將在進階語法中詳細説明。

1. 比較運算子

既然布林類型用來判斷條件是否成立，那就不得不提一下 Python 中的**比較運算子（Comparison Operators）**。比較運算子包含等於 ==、不等於 !=、大於 >、小於 <、大於等於 >=、小於或等於 <=。**比較運算子和 +、-、*、//、** 這些算術運算子（在 2.3.2 節中提到過）最大的區別是：前者用來判**

斷符號左右兩邊的變數和資料是否滿足運算子本身的條件，並且傳回值是布林值，後者則單純做加減乘除等運算，傳回值是整數或浮點數。

在編輯器模式下，使用比較運算子後可以馬上看到傳回的布林值 True 或 False。如果是指令稿模式，則需要配合 print 指令才能看到，舉例如下。

```
>>> a = 100
>>> a == 100
True
>>> a == 1000
False
>>> a != 1000
True
>>> a != 100
False
>>> a > 99
True
>>> a > 101
False
>>> a < 101
True
>>> a < 99
False
>>> a >= 99
True
>>> a >= 101
False
>>> a <= 100
True
>>> a <= 99
False
>>>
```

2. 邏輯運算子

除了比較運算子,使用**邏輯運算子**(Logical Operators)也能傳回布林值。邏輯運算子有 3 種:與(and)、或(or)、非(not)。學過離散數學的讀者對與、或、非的邏輯運算不會感到陌生,在邏輯運算中使用的真值表(Truth Table)如下。

真值表(Truth Table)				
P	Q	P 與 Q(and)	P 或 Q(or)	非 P(not)
True	True	True	True	False
True	False	False	True	False
False	True	False	True	True
False	False	False	False	True

邏輯運算子在 Python 中的使用舉例如下。

```
>>> A = True
>>> B = True
>>> A and B
True
>>> A or B
True
>>> not A
False
>>>
>>>
>>> A = False
>>> B = True
>>> A and B
False
>>> A or B
True
>>> not A
```

```
True
>>>
>>> A = False
>>> B = False
>>> A and B
False
>>> A or B
False
>>> not A
True
>>>
```

2.4.6 集合、元組、空值

作為同樣需要網路工程師掌握的 Python 資料類型，**集合**（Set）、**元組**（Tuple）、**空值**（None）相對來說使用頻率不如字串、整數、浮點數、串列、字典及布林類型那麼高，這裡進行簡單介紹。

1. 集合

- 集合是一種特殊的串列，裡面**沒有重複的元素**，因為每個元素在集合中都只有一個，所以集合沒有 count() 方法。
- 集合可以透過大括號 {}（與字典一樣，但是集合沒有鍵值對）或 set() 函數建立。

```
>>> interfaces = { 'Fa0/0' , 'Fa0/1', 'Fa0/2'}
>>> type(interfaces)
<class 'set'>
>>> vendors = set(['Cisco','Juniper','Arista','Cisco'])
>>> type(vendors)
<class 'set'>
>>> print (vendors)
{'Cisco', 'Arista', 'Juniper'}
>>>
```

vendors 串列中有兩個重複的元素，即 "Cisco"，在用 set() 函數將它轉換成集合後，多餘的 "Cisco" 被去掉，只保留一個。

■ 集合是無序的，不能像串列那樣使用索引值，也不具備 index() 函數。

```
>>> vendors[2]
Traceback (most recent call last):
  File "<stdin>", line 1, in <module>
TypeError: 'set' object is not subscriptable
>>>
>>> vendors.index('Cisco')
Traceback (most recent call last):
  File "<stdin>", line 1, in <module>
AttributeError: 'set' object has no attribute 'index'
>>>
```

下面介紹與集合有關的方法和函數。

❏ add()

add() 用來在一組集合增加新元素，**其傳回值依然是集合**，舉例如下。

```
>>> vendors.add('Huawei')
>>> vendors
{'Huawei'，'Cisco'，'Arista'，'Juniper'}
>>>
```

❏ remove()

remove() 用來刪除一組集合中已有的元素，**其傳回值依然是集合**，舉例如下。

```
>>> vendors.remove('Arista')
>>> vendors
{'Huawei'，'Cisco'，'Juniper'}
>>>
```

2. 元組

- 與集合一樣，元組也是一種特殊的串列。它與串列最大的區別是：可以任意對串列中的元素進行增添、刪除、修改，而元組則不可以。一旦建立元組，將無法對其做任何形式的更改，所以元組沒有 append()、insert()、pop()、add() 和 remove()，只保留了 index() 和 count() 兩種方法。
- 元組可以透過小括號 () 建立，也可以使用 tuple() 函數建立。
- 與串列一樣，元組是有序的，可以對元素進行索引。

```
>>> vendors = ('Cisco','Juniper','Arista')
>>> print (vendors)
('Cisco', 'Juniper', 'Arista')
>>> print (vendors[1])
Juniper
>>> vendors[2] = 'Huawei'
Traceback (most recent call last):
 File "<stdin>", line 1, in <module>
TypeError: 'tuple' object does not support item assignment #不能對元組做任何
形式的修改
>>>
```

下面介紹與元組有關的方法和函數。

❑ index()

元組的 index() 與串列用法相同，都用來查詢指定元素的索引值。index() 的傳回值為整數，舉例如下。

```
>>> vendors
('Cisco', 'Juniper', 'Arista')
>>> vendors.index('Cisco')
0
>>>
```

❑ count()

元組的 count() 與串列用法相同，都用來查詢指定元素在元組中的數量。
count() 的傳回值為整數，舉例如下。

```
>>> vendors = ('Cisco','Juniper','Arista','Cisco')
>>> vendors.count('Cisco')
2
>>> vendors.count('Juniper')
1
>>>
```

3. 空值

空值是比較特殊的資料類型，它沒有附帶的函數和方法，也無法做任何算
術和邏輯運算，但是可以被設定值給一個變數，舉例如下。

```
>>> type(None)
<type 'NoneType'>
>>> None == 100
False
>>> a = None
>>> print (a)
None
```

空值（None）較常用在判斷敘述和正規表示法中。對網路工程師來説，日
常工作中需要經常使用顯示指令（show 或 display）來對網路裝置進行校
正或查詢網路資訊，通常這種顯示指令都會列出很多回應內容，而大多時
候我們只需要關注其中的一兩項參數即可。如果用 Python 來實現網路運行
維護自動化，則需要使用正規表示法來告訴 Python 應該抓取哪一個「關鍵
字」（即我們想要的參數）；而空值則可以用來判斷「關鍵字」抓取得是否
成功。關於判斷敘述和正規表示法的用法將在第 3 章中詳細介紹。

Python 進階語法

前兩章分別介紹了 Python 在 Windows 和 Linux 裡的安裝和使用方法，詳細說明了網路工程師需要掌握的 Python 資料類型及每種資料類型附帶的函數和方法的用法。本章將說明 Python 中的條件（判斷）敘述、循環敘述、文字檔的讀寫、自訂函數、模組、正規表示法及異常處理等網路工程師需要掌握的 Python 進階語法知識。

3.1 條件（判斷）敘述

在 Python 中，條件陳述式（Conditional Statements）又叫作判斷敘述，判斷敘述由 if、elif 和 else 3 種敘述組成，其中 if 為強制敘述，可以獨立使用，elif 和 else 為可選敘述，並且不能獨立使用。判斷敘述配合布林值，透過判斷一條或多行敘述的條件是否成立（True 或 False），進一步決定下一步的動作，如果判斷條件成立（True），則執行 if 或 elif 敘述下的程式；如果判斷條件不成立（False），則執行 else 敘述下的程式；如果沒有 else 敘述，則不做任何事情。

布林值是判斷敘述不可或缺的部分，在基本語法中講到的比較運算子、邏輯運算子，以及字串附帶的 startswith()、endswith()、isdigit()、isalpha() 等方法，還有下面將講到的成員運算子等都會傳回布林值。下面就舉例說明它們各自在 Python 判斷敘述中的應用場景。

3.1.1 透過比較運算子作判斷

在講布林類型時，我們已經提到與布林值 True 和 False 息息相關的各種比較運算子，包含等於 ==、不等於 !=、大於 >、小於 <、大於等於 >= 和小於或等於符號 <=，因為使用比較運算子後會直接傳回布林值，所以比較運算子在判斷敘述中會經常被用到，舉例如下。

首先我們用指令稿模式寫一段程式。

```
[root@CentOS-Python ~]# cat lab.py

# coding=utf-8
final_score = input('請輸入你的CCNA考試分數:')
if int(final_score) > 811:
    print ('恭喜你考試及格。')
elif int(final_score) == 811:
    print ('恭喜你壓線考試及格。')
else:
    print ('成績不及格。')
```

這段程式用來讓使用者輸入自己的 CCNA 考試成績並作判斷，假設及格線為 811 分，如果使用者所得分數大於 811 分，則列印「恭喜你考試及格。」；如果分數剛剛等於 811 分，則列印「恭喜你壓線考試及格。」；如果低於 811 分，則列印「成績不及格。」。這段程式需要注意以下幾點。

- 寫在 if、elif 和 else 下的程式都做了**程式縮排**（Indentation），也就是 print() 函數的前面保留了 4 個空格。不同於 C、C++、Java 等語言，

Python 要求嚴格的程式縮排，目的是讓程式工整並且具有可讀性，方便閱讀和修改。縮排不一定必須是 4 個空格，兩個空格或 8 個空格都是允許的，目前最常見的是 4 個空格的縮排。

■ if、elif 和 else 敘述的結尾必須接冒號，這點需要注意。

■ 使用 input() 函數讓使用者輸入自己的分數，並把它設定值給 final_score 變數。在第 2 章裡講過，input() 函數的傳回值是字串，因為要與 811 這個整數做比較，所以需要透過 int() 函數先將 final_score 從字串轉為整數。

■ 與 if 和 elif 敘述不同，else 後面不需要再給任何判斷條件。

執行這段指令稿看效果。

```
[root@CentOS-Python ~]# python3.8 lab.py
請輸入你的CCNA考試分數:1000
恭喜你考試及格。
[root@CentOS-Python ~]# python3.8 lab.py
請輸入你的CCNA考試分數:811
恭喜你壓線考試及格。
[root@CentOS-Python ~]# python3.8 lab.py
請輸入你的CCNA考試分數:700
成績不及格。
[root@CentOS-Python ~]#
```

3.1.2 透過字串方法 + 邏輯運算子作判斷

當使用 input() 函數讓使用者輸入內容時，你無法保障使用者輸入的內容合乎標準。例如你給使用者 6 個選項，每個選項分別對應一個動態路由式通訊協定的名稱（選項 1：RIP；選項 2：IGRP；選項 3：EIGRP；選項 4：OSPF；選項 5：ISIS；選項 6：BGP），提示使用者輸入路由式通訊協定的選項號碼來查詢該路由式通訊協定的類型，然後讓 Python 根據使用者輸入

的選項號碼告訴使用者該路由式通訊協定屬於鏈路狀態路由式通訊協定、
距離向量路由式通訊協定，還是路徑適量路由式通訊協定。

**這裡你無法保障使用者輸入的一定是整數，即讓使用者輸入的是整數，也
無法保障輸入的是 1 ～ 6 的數字。** 因為 input() 函數的傳回值是字串，所以
可以首先使用字串的 isdigit() 函數來判斷使用者輸入的內容是否為整數，
這是判斷條件之一。然後透過 int() 將該字串數字轉換成整數，繼續判斷該
整數是否介於 1 和 6 之間（包含 1 和 6），這是判斷條件之二。再將這兩個
判斷條件透過邏輯運算子 and 來判斷他們是否同時成立。如果成立，則傳
回對應的答案；如果不成立，則提示使用者輸入的內容不符合標準並終止
程式。程式如下。

```
[root@CentOS-Python ~]# cat lab.py

# coding=utf-8

print ('''請根據對應的號碼選擇一個路由式通訊協定:
1. RIP
2. IGRP
3. EIGRP
4. OSPF
5. ISIS
6. BGP ''')

option = input('請輸入你的選項(數字1-6): ')
if option.isdigit() and 1 <= int(option) <= 6:
    if option == '1' or option == '2' or option == '3':
        print ('該路由式通訊協定屬於距離向量路由式通訊協定。')
    elif option == '4' or option == '5':
        print ('該路由式通訊協定屬於鏈路狀態路由式通訊協定。')
    else:
        print ('該路由式通訊協定屬於路徑向量路由式通訊協定。')
else:
    print ('選項無效，程式終止。')
```

這裡我們用到了**巢狀結構 if 敘述**。在某一個條件成立（判斷為 True）後，如果還需要檢查其他子條件，就可以用巢狀結構 if 敘述來完成。在巢狀結構 if 敘述中，一組 if、elif 和 else 可以建置在另一組 if、elif 和 else 中，不過需要注意縮排。

執行程式，測試效果如下。

```
[root@CentOS-Python ~]# python3.8 lab.py

請根據對應的號碼選擇一個路由式通訊協定:
1. RIP
2. IGRP
3. EIGRP
4. OSPF
5. ISIS
6. BGP
請輸入你的選項(數字1-6):1
該路由式通訊協定屬於距離向量路由式通訊協定。

[root@CentOS-Python ~]# python lab.py
請根據對應的號碼選擇一個路由式通訊協定:
1. RIP
2. IGRP
3. EIGRP
4. OSPF
5. ISIS
6. BGP
請輸入你的選項(數字1-6):4
該路由式通訊協定屬於鏈路狀態路由式通訊協定。

[root@CentOS-Python ~]# python lab.py
請根據對應的號碼選擇一個路由式通訊協定:
1. RIP
2. IGRP
3. EIGRP
```

```
4. OSPF

5. ISIS

6. BGP
請輸入你的選項(數字1-6)：6
該路由式通訊協定屬於路徑向量路由式通訊協定。


[root@CentOS-Python ~]# python lab.py
請根據對應的號碼選擇一個路由式通訊協定：
1. RIP

2. IGRP

3. EIGRP

4. OSPF

5. ISIS

6. BGP
請輸入你的選項(數字1-6)：abc
選項無效，程式終止。


[root@CentOS-Python ~]# python lab.py
請根據對應的號碼選擇一個路由式通訊協定：
1. RIP

2. IGRP

3. EIGRP

4. OSPF

5. ISIS

6. BGP
請輸入你的選項(數字1-6)：8
選項無效，程式終止。
[root@CentOS-Python ~]#
```

3.1.3 透過成員運算子作判斷

成員運算子用於判斷是否可以在指定的一組字串、串列、字典和元組中找到一個指定的值或變數，如果能找到，則傳回布林值 True；如果找不到，則傳回布林值 False。成員運算子有兩種：in 和 not in，舉例如下。

```
>>> netdevops = '網路工程師需要學習Python嗎？'
>>> 'Python' in netdevops
True
>>> 'Java' in netdevops
False
>>> 'C++' not in netdevops
True
>>>
>>> interfaces = ['Gi1/1','Gi1/2','Gi1/3','Gi1/4','Gi1/5']
>>> 'Gi1/1' in interfaces
True
>>> 'Gi1/10' in interfaces
False
>>> 'Gi1/3' not in interfaces
False
>>>
```

依靠成員運算子，我們還可以將類似上一節的指令稿簡化。上一節的指令稿列出的選項只有 6 種，我們尚且能夠使用 if option == '1' or option == '2' or option == '3': 這種的方法來列舉所需的選項，但是如果所需選項超過 20 個甚至上百個，那麼再使用這種方法豈不是太笨了？這時可以使用第 2 章講的 range() 函數配合 list() 函數創造一個整數串列，然後配合成員運算子來判斷使用者輸入的選項號碼是否存在於該整數串列中。

按照這種想法，我們將上一節的指令稿做以下簡化。

```
[root@CentOS-Python ~]# cat lab.py
# coding=utf-8

print ('''請根據對應的號碼選擇一個路由式通訊協定:
1. RIP
2. IGRP
3. EIGRP
4. OSPF
5. ISIS
```

```
6. BGP ''')

option = input('請輸入你的選項(數字1-6): ')
if option.isdigit() and int(option) in list(range(1, 7)):
    if int(option) in list(range(1, 4)):
        print ('該路由式通訊協定屬於距離向量路由式通訊協定。')
    elif int(option) in list(range(4, 6)):
        print ('該路由式通訊協定屬於鏈路狀態路由式通訊協定。')
    else:
        print ('該路由式通訊協定屬於路徑向量路由式通訊協定。')
else:
    print ('選項無效, 程式終止。')

[root@CentOS-Python ~]#
```

之前依靠比較運算子＋邏輯運算子作判斷的 if option == '1' or option == '2' or option == '3': 方法已經被成員運算子 +range() 函數簡化為 if int(option) in list(range(1, 4)): 了。同理，判斷選項在整數 1～6 之間也已經透過成員運算子配合 range() 和 list() 函數寫成 int(option) inrange(1, 7) 了。需要注意的是，由於 input() 函數傳回的是字串，因此需要把變數 option 先透過 int() 函數轉為整數才能使用成員運算子 in 來判斷它是否存在於 range() 和 list() 函數所建立的整數串列中。

3.2 循環敘述
. . . .

Python 中最常用的循環敘述（Looping Statements）有兩種：while 和 for。除此之外，還有檔案反覆運算器（File Iterator）、串列解析式（List Comprehension）等循環工具，不過對網路工程師來說，用得最多的還是 while 和 for，因此本節將只說明這兩種循環敘述。

3.2.1 while 敘述

在 Python 中，while 敘述用於循環執行一段程式，它和 if 敘述一樣，兩者都離不開判斷敘述和縮排。每當寫在 while 敘述下的程式被執行一次，程式就會自動回到「頂上」(也就是 while 敘述的開頭部分)，根據 while 後的判斷敘述的傳回值來決定是否要再次執行該程式，如果判斷敘述的傳回值為 True，則繼續執行該程式，一旦判斷敘述的傳回值為 False，則該 while 循環隨即終止，如此反覆。如果需要中途強行中止 while 循環，則需要使用 break 敘述。下面透過 3 個實例幫助大家了解。

❑ 例 1

```
>>> a = 1
>>> b = 10
>>> while a < b:
...     print (a)
...     a += 1
...
1
2
3
4
5
6
7
8
9
>>>
```

上面程式中，我們用 while 循環來判斷變數 a 是否小於 b，如果判斷結果為 True，則列印出變數 a 的值，並且每次都讓 a 的值加 1，如此反覆循環，直到第 10 次執行該 while 循環，a = 10，因為 a < b 不再成立 (10<10 不成立)，程式隨即終止。

❏ 例 2

```
>>> vendors = ['Cisco','Huawei','Juniper','Arista','HPE','Extreme']
>>> while len(vendors) > 0:
...     vendors.pop()
...     print (vendors)
...
'Extreme'
['Cisco', 'Huawei', 'Juniper', 'Arista', 'HPE']
'HPE'
['Cisco', 'Huawei', 'Juniper', 'Arista']
'Arista'
['Cisco', 'Huawei', 'Juniper']
'Juniper'
['Cisco', 'Huawei']
'Huawei'
['Cisco']
'Cisco'
[]
>>>
```

上面程式中，我們用 while 循環配合 len() 函數來判斷串列 vendors 的長度是否大於 0，如果判斷結果為 True，則用 pop() 方法從串列中刪掉一個元素，並且隨即列印串列裡剩餘的元素。最後當串列中所有的元素都被移除時，串列的長度為 0，while len(vendors)>0: 的傳回值為 False，該 while 循環也就隨即終止。

❏ 例 3

在 3.1.2 節和 3.1.3 節的案例程式中，一旦使用者輸入的選項不符合標準，程式就會立即中止，使用者必須再次手動執行一次指令稿重新輸入選項，這樣顯得很笨拙。借助 while 循環，可以不斷地重複執行 input() 函數，直到使用者輸入正確的選項號碼。最佳化後的指令稿程式如下。

```
# coding=utf-8

print ('''請根據對應的號碼選擇一個路由式通訊協定:
1. RIP
2. IGRP
3. EIGRP
4. OSPF
5. ISIS
6. BGP ''')

while True:
    option = input('請輸入你的選項(數字1-6): ')
    if option.isdigit() and int(option) in list(range(1,7)):
        if int(option) in list(range(1,4)):
            print ('該路由式通訊協定屬於距離向量路由式通訊協定。')
        elif int(option) in list(range(5,7)):
            print ('該路由式通訊協定屬於鏈路狀態路由式通訊協定。')
        else:
            print ('該路由式通訊協定屬於路徑向量路由式通訊協定。')
        break
    else:
        print ('選項無效,請再次輸入。')

[root@CentOS-Python ~]#
```

這裡我們使用了 while True。while True 是一種很常見的 while 循環的用法,因為這裡的判斷條件的結果已經手動指定了 True,表示判斷條件將永久成立,也就表示 while 下面的程式將被無數次重複執行,進一步引起「無限循環」(Indefinite Loop)的問題。為了避免無限循環,我們必須在程式碼中使用 break 敘述來終止 while 循環,注意 break 在上面程式裡的位置,帶著這個問題去思考,你會更加明白縮排在 Python 中的重要性。

執行程式看效果。

```
[root@CentOS-Python ~]#python lab.py
請根據對應的號碼選擇一個路由式通訊協定:
1. RIP
2. IGRP
3. EIGRP
4. OSPF
5. ISIS
6. BGP
請輸入你的選項(數字1-6): 7
選項無效,請再次輸入。
請輸入你的選項(數字1-6): a
選項無效,請再次輸入。
請輸入你的選項(數字1-6): ..!!!
選項無效,請再次輸入。
請輸入你的選項(數字1-6): 134ui134lkadjl
選項無效,請再次輸入。
請輸入你的選項(數字1-6): 1
該路由式通訊協定屬於距離向量路由式通訊協定。

[root@CentOS-Python ~]#
```

3.2.2 for 敘述

同為循環敘述,for 敘述的循環機制和 while 敘述完全不同:while 敘述需要配合判斷敘述來決定什麼時候開始循環和中止循環,而 for 敘述則用來檢查一組可反覆運算的序列,可反覆運算的序列包含字串、串列、元組等。在將這些序列中的元素檢查完後,for 敘述的循環也隨即終止。for 敘述的基本語法格式如下。

```
for item in sequence:
    statements
```

這裡的 sequence 為可反覆運算的序列（如字串、串列、元組），而 item 可以視為該序列裡的每個元素（item 名稱可以任意選取），statements 則是循環本體（將要循環的程式部分）。

舉例如下。

❑ 例 1

```
>>> for letter in 'Python':
...  print (letter)
...
P
y
t
h
o
n
>>>
```

我們用 letter 作為 for 敘述中的 item 來檢查字串 "Python"，並將該字串中的元素依次全部列印出來，獲得 P、y、t、h、o、n。

❑ 例 2

```
>>> sum = 0
>>> for number in range(1，6):
...     sum = sum + number
...     print (sum)
...
1
3
6
10
15
>>>
```

我們將 0 設定值給變數 sum，然後用 number 作為 for 敘述中的 item 來檢查 range(1,6)，傳回 1 ～ 5 的 5 個整數，將這 5 個整數依次與 sum 累加，每累加一次用 print() 函數列印出結果，最後獲得 1、3、6、10、15。

❑ 例 3

```
>>> routing_protocols = ['RIP','IGRP','EIGRP','OSPF','ISIS','BGP']
>>> link_state_protocols = ['OSPF','ISIS']
>>> for protocols in routing_protocols:
...     if protocols not in link_state_protocols:
...         print (protocols + '不屬於鏈路狀態路由式通訊協定：' )
RIP不屬於鏈路狀態路由式通訊協定：
IGRP不屬於鏈路狀態路由式通訊協定：
EIGRP不屬於鏈路狀態路由式通訊協定：
BGP不屬於鏈路狀態路由式通訊協定：
>>>
```

我們分別建立兩個串列：routing_protocols 和 link_state_protocols，串列中的元素是對應的路由式通訊協定和鏈路狀態路由式通訊協定。首先用 protocols 作為 for 敘述中的 item 來檢查第一個串列 routing_protocols，然後使用 if 敘述來判斷哪些 protocols 不屬於第二個串列 link_state_protocols 中的元素，並將它們列印出來。

正如前面講到的，上述三個實例中寫在 for 後面的 letter、number 和 protocols 代表將要檢查的可反覆運算序列裡的每一個元素（即 item 名稱），**它們的名稱可以由使用者隨意制定**，例如在例 1 中，我們把 letter 換成 a 也沒問題。

```
>>> for a in 'Python':
...   print (a)
...
P
y
```

```
t
h
o
n
```

通常建議取便於了解的 item 名稱，像 a 這種的 item 名稱在做實驗或練習時可以偷懶使用，在實際的工作程式中用這種毫無意義的 item 名稱一定是會被人詬病的。

3.3 文字檔的讀／寫

在日常網路運行維護中，網路工程師免不了要和大量的文字檔進行處理，例如用來批次設定網路裝置的指令範本檔案，儲存所有網路裝置 IP 位址的檔案，以及備份網路裝置 show run 輸出結果之類的設定備份檔案。正因如此，知道如何使用 Python 來存取和管理文字檔是學習網路運行維護自動化技術的網路工程師必須掌握的一項 Python 基礎知識。

3.3.1 open() 函數及其模式

在 Python 中，我們可以透過 open() 函數來存取和管理文字檔，open() 函數用來開啟一個文字檔並建立一個檔案物件（File Object），透過檔案物件附帶的多種函數和方法，可以對文字檔執行一系列存取和管理操作。在說明這些函數和方法之前，首先建立一個名為 test.txt 的測試文字檔，該檔案包含 5 個網路裝置廠商的名字，內容如下。

```
[root@CentOS-Python ~]# cat test.txt
Cisco
Juniper
Arista
```

```
H3C
Huawei
[root@CentOS-Python ～]#
```

然後用 open() 函數存取該檔案。

```
>>> file = open('test.txt','r')
```

我們透過 open() 函數的 r 模式（唯讀）存取 test.txt 檔案，並傳回一個檔案物件，再將該檔案物件設定值給 file 變數。r（reading）**是預設存取模式**，除此之外，open() 函數還有很多其他檔案存取模式。這裡只介紹網路工程師最常用的幾種模式，如下表所示。

模式	作　用
r	以唯讀方式開啟檔案，r 模式只能開啟已存在的檔案。如果檔案不存在，則會顯示出錯
w	開啟檔案並只用於寫入。如果檔案已經存在，則原有內容將被刪除覆蓋；如果檔案不存在，則建立新檔案
a	以追加方式開啟檔案。如果檔案已經存在，則原有內容不會被刪除覆蓋，新內容將增加在原有內容後面；如果檔案不存在，則建立新檔案
r+	以讀寫方式開啟檔案，r+ 模式只能開啟已存在的檔案。如果檔案不存在，則會顯示出錯
·w+	以讀寫方式開啟檔案。如果檔案已經存在，則原有內容將被刪除覆蓋；如果檔案不存在，則建立新檔案
a+	以讀寫方式開啟檔案。如果檔案已經存在，則原有內容不會被刪除覆蓋，新內容將增加在原有內容後面；如果檔案不存在，則建立新檔案

網路工程師必須熟練掌握 open() 函數的上述 6 種模式，關於它們的實際使用將在下一節中舉例說明。

3.3.2 檔案讀取

在使用 open() 函數建立檔案物件之後，我們並不能馬上讀取檔案裡的內容。如下所示，在建立了檔案物件並將它設定值給 file 變數後，如果用 print() 函數將 file 變數列印出來，則只會獲得檔案名稱、open() 函數的存取模式及該檔案物件在記憶體中的位置（0x7fa194215660）等資訊。

```
>>> file = open('test.txt','r')
>>> print (file)
<open file 'test.txt', mode 'r' at 0x7fa194215660>
>>>
```

要想讀取檔案裡的實際內容，我們還需要用 read()、readline() 或 readlines() 3 種方法中的一種。因為這 3 種方法都和讀取有關，因此 open() 函數中只允許寫入的 w 模式和只允許追加的 a 模式不支援它們，而其他 4 種模式則都沒有問題，舉例如下。

```
#w模式不支援read()，readline()，readlines()
>>> file = open('test.txt','w')
>>> print (file.read())
Traceback (most recent call last):
  File "<stdin>", line 1, in <module>
io.UnsupportedOperation: not readable
>>>

#a模式也不支援read()，readline()，readlines()
>>> file = open('test.txt','a')
>>> print (file.readline())
Traceback (most recent call last):
  File "<stdin>", line 1, in <module>
io.UnsupportedOperation: not readable
>>>
```

read()、readline() 和 readlines() 是學習 open() 函數的重點內容，三者的用法和差異很大，其中 readlines() 更是重中之重（原因後面會講到），網路工程師必須熟練掌握。下面對這 3 種函數一一說明。

1. read()

read() 方法讀取文字檔裡的全部內容，**傳回值為字串**。

```
>>> file = open('test.txt')
>>> print (file.read())
Cisco
Juniper
Arista
H3C
Huawei

>>> print (file.read())
>>>
```

我們嘗試連續兩次列印 test.txt 檔案的內容，第一次列印出的內容沒有任何問題，為什麼第二次列印的時候內容為空了呢？這是因為在使用 read() 方法後，**檔案指標的位置從檔案的開頭移動到了尾端**，要想讓檔案指標回到開頭，必須使用 seek() 函數，方法如下。

```
>>> file.seek(0)
0
>>> file.tell()
0
>>> print (file.read())
Cisco
Juniper
Arista
H3C
Huawei
```

```
>>> file.tell()
32
>>>
```

我們用 seek(0) 將檔案指標從尾端移回開頭，並且用 tell() 方法確認檔案指標的位置（檔案開頭的位置為 0），隨後使用 read() 方法列印檔案內容並成功，之後再次使用 tell() 方法確認檔案指標的位置，可以發現指標現在已經來到檔案尾端處（32）。這個 32 是怎麼得來的？下面我們去掉 print() 函數，再次透過 read() 方法來讀取一次檔案內容。

```
>>> file.seek(0)
0
>>> file.read()
'Cisco\nJuniper\nArista\nH3C\nHuawei\n'
>>>
```

去掉 print() 函數的目的是能清楚地看到分行符號 \n，如果這時從左往右數，則會發現 Cisco(5) + \n(1) + Juniper(7) + \n(1) + Arista(6) + \n(1) + H3C(3) + \n(1) + Huawei(6) + \n(1) = 32，這就解釋了為什麼在檔案指標移動到檔案尾端後，tell() 方法傳回的檔案指標的位置是 32。檔案指標的位置及 seek() 和 tell() 方法的用法是文字檔存取和管理中很重要但又容易被忽略的基礎知識，網路工程師務必熟練掌握。

2. readline()

readline() 與 read() 的區別是它不會像 read() 那樣把文字檔的所有內容一次性都讀完，而是會一排一排地去讀。**readline() 的傳回值也是字串**。舉例如下。

```
>>> file = open('test.txt')
>>> print (file.readline())
Cisco
```

```
>>> print (file.readline())
Juniper

>>> print (file.readline())
Arista

>>> print (file.readline())
H3C

>>> print (file.readline())
Huawei

>>> print (file.readline())

>>>
```

readline() 方法每次傳回檔案的一排內容,順序由上至下 (這裡出現的空排部分是因為分行符號的緣故)。另外,檔案指標會跟隨移動直到檔案尾端,因此最後一個 print (file.readline()) 的傳回值為空。

3. readlines()

readlines() 與前兩者最大的區別是它的傳回值不再是字串,而是串列。可以説 readlines() 是 read() 和 readline() 的結合體,首先它同 read() 一樣把文字檔的所有內容都讀完。另外,它會像 readline() 那樣一排一排地去讀,並將每排的內容以串列元素的形式傳回,舉例如下。

```
>>> file = open('test.txt')
>>> print (file.readlines())
['Cisco\n', 'Juniper\n', 'Arista\n', 'H3C\n', 'Huawei\n']
>>> file.seek(0)
>>> devices = file.readlines()
>>> print (devices[0])
Cisco
```

```
>>> print (devices[1])
Juniper

>>> print (devices[2])
Arista

>>> print (devices[3])
H3C

>>> print (devices[4])
Huawei

>>>
```

同 read() 和 readline() 一樣，使用一次 readlines() 後，檔案指標會移動到檔案的尾端。為了避免每次重複使用 seek(0) 的麻煩，可以將 readlines() 傳回的串列設定值給一個變數，即 devices，之後便可以使用索引值來一個一個地驗證串列裡的元素。注意 readlines() 傳回的串列裡的元素都帶分行符號 \n，這與我們手動建立的普通串列是有區別的。

同 read() 和 readline() 相比，筆者認為 readlines() 應該是網路運行維護中使用頻率最高的一種讀取文字檔內容的方法，因為它的傳回值是串列，透過串列我們可以做很多事情。舉個實例，現在有一個名為 ip.txt 的文字檔，該檔案儲存了一大堆沒有規律可循的交換機的管理 IP 位址，實際如下。

```
[root@CentOS-Python ~]# cat ip.txt
172.16.100.1
172.16.30.1
172.16.41.1
172.16.10.1
172.16.8.1
172.16.112.1
172.16.39.1
```

```
172.16.121.1
172.16.92.1
172.16.73.1
192.168.54.1
192.168.32.1
192.168.2.1
10.3.2.1
10.58.23.3
192.168.230.29
10.235.21.42
192.168.32.32
10.4.3.3
172.16.30.2
172.16.22.30
172.16.111.33
[root@CentOS-Python ~]#
```

現在需要回答 3 個問題：

（1）怎麼使用 Python 來確定該檔案有多少個 IP 位址？

（2）怎麼使用 Python 來找出該檔案中的 B 類別 IP 位址（172.16 開頭的 IP 位址），並將它們列印出來？

（3）怎麼使用 Paramiko 來批次 SSH 登入這些交換機修改或檢視設定（後面會講到 Paramiko）？

答案（1）：最好的方法是使用 open() 函數的 readlines() 來讀取該檔案的內容，因為 readlines() 的傳回值是串列，可以方便我們使用 len() 函數來判斷該串列的長度，進一步獲得該檔案中所包含 IP 位址的數量，舉例如下。

```
>>> f = open('ip.txt')
>>> print (len(f.readlines()))
22
>>>
```

僅透過兩行程式，就獲得了結果：22 個 IP 位址，readlines() 傳回的串列是
不是很方便？

答案（2）：最好的方法依然是使用 readlines() 來完成，因為 readlines() 傳
回的串列中的元素的資料類型是字串，可以使用 for 循環來檢查所有的字
串元素，然後配合 if 敘述，透過字串的 startswith() 函數判斷這些 IP 位址
是否以 172.16 開頭。如果是，則將它們列印出來，舉例如下。

```
>>> f = open('ip.txt')
>>> for ip in f.readlines():
...     if ip.startswith('172.16'):
...          print (ip)
...
172.16.100.1

172.16.200.1

172.16.130.1

172.16.10.1

172.16.8.1

172.16.112.1

172.16.39.1

172.16.121.1

172.16.92.1

172.16.73.1

172.16.30.2
```

```
172.16.22.30

172.16.111.33
>>>
```

需要注意的是，因為 readlines() 傳回的串列中的元素是帶分行符號 \n 的，所以列印出來的每個 B 類別 IP 位址之間都空了一排，影響閱讀和美觀，解決方法也很簡單，只需要使用第 2 章講過的 strip() 函數去掉分行符號即可，舉例如下。

```
>>> f = open('ip.txt')
>>> for ip in f.readlines():
...     if ip.startswith('172.16'):
...             print (ip.strip())
...
172.16.100.1
172.16.200.1
172.16.130.1
172.16.10.1
172.16.8.1
172.16.112.1
172.16.39.1
172.16.121.1
172.16.92.1
172.16.73.1
172.16.30.2
172.16.22.30
172.16.111.33
>>>
```

答案（3）：同答案（2）一樣，因為 readlines() 傳回的是串列，我們可以透過 for 循環來一一存取該串列中的每個元素，也就是交換機的每個 IP 位址，進而達到使用 Paramiko 來一一登入每個交換機做設定的目的，該指令稿的程式如下。

```
import Paramiko

username = input('Username: ')
password = input('Password: ')
f = open('ip.txt')

for ip in f.readlines():
    ssh_client = Paramiko.SSHClient()
    ssh_client.set_missing_host_key_policy(Paramiko.AutoAddPolicy())
    ssh_client.connect(hostname=ip,username=username,password=password)
    print ("Successfully connect to ", ip)
```

- 關於 Paramiko 的用法在第 4 章會講到，這裡看不懂沒關係。
- 這裡透過使用 for 循環配合 readlines() 傳回串列的方式來存取 ip.txt 檔案中的交換機管理 IP 位址。
- 該 for 循環會嘗試一一登入 ip.txt 的所有交換機 IP 位址，每成功登入一個交換機都隨即列印資訊「"Successfully connect to"，ip」來提醒使用者登入成功。

open() 函數的 readlines() 在網路運行維護中的用處遠不止這 3 點，這裡只是列出了幾個比較典型的實用實例。再次強調，每一位學習 Python 的網路工程師都必須熟練掌握它。

3.3.3 檔案寫入

在使用 open() 函數建立檔案物件後，我們可以使用 **write() 函數**來對檔案寫入資料。顧名思義，既然 write() 函數與檔案寫入相關，**那麼只允許唯讀的 r 模式並不支援它**，而其他 5 種模式則不受限制（包含 r+ 模式），舉例如下。

```
>>> f = open('test.txt','r')
>>> f.write()
```

```
Traceback (most recent call last):
  File "<stdin>", line 1, in <module>
TypeError: write() takes exactly one argument (0 given)
>>>
```

write() 函數在 r+、w/w+、a/a+ 這 5 種模式中的應用說明如下。

1. r+

在 r+ 模式下使用 write() 函數，新內容會增加在檔案的開頭部分，而且會覆蓋開頭部分原來已有的內容，舉例如下。

```
#文字修改前的內容
[root@CentOS-Python ~]# cat test.txt
Cisco
Juniper
Arista
H3C
Huawei

[root@CentOS-Python ~]#

#在r+模式下使用write()函數修改文字內容
>>> f = open('test.txt','r+')
>>> f.write('Avaya')
>>> f.close()

#文字修改後的內容
[root@CentOS-Python ~]# cat test.txt
Avaya
Juniper
Arista
H3C
Huawei

[root@CentOS-Python ~]#
```

可以看到，文字開頭的 Cisco 已經被 Avaya 覆蓋了。這裡注意使用 write()
函數對文字寫入新內容後，必須再用 close() 方法將文字關閉，這樣新寫入
的內容才能被儲存。

2. w/w+

在 w/w+ 模式下使用 write() 函數，新內容會增加在檔案的開頭部分，已存
在的檔案的內容將完全被清空，舉例如下。

```
#文字修改前的內容
[root@CentOS-Python ～]# cat test.txt
Avaya
Juniper
Arista
H3C
Huawei

[root@CentOS-Python ～]#

#在w模式下使用write()函數修改文字內容
>>> f = open('test.txt','w')
>>> f.write('test')
>>> f.close()

#文字修改後的內容
[root@CentOS-Python ～]# cat test.txt
test

[root@CentOS-Python ～]#

#在w+模式下使用write()函數修改文字內容
>>> f = open('test.txt','w+')
>>> f.write('''Cisco
... Juniper
... Arista
```

```
...  H3C
...  Huawei\n''')
>>> f.close()

#文字修改後的內容
[root@CentOS-Python ~]# cat test.txt
Cisco
Juniper
Arista
H3C
Huawei

[root@CentOS-Python ~]#
```

3. a/a+

在 a/a+ 模式下使用 write() 函數，新內容會增加在檔案的尾端部分，已存
在的檔案的內容將不會被清空，舉例如下。

```
#文字修改前的內容：
[root@CentOS-Python ~]# cat test.txt
Cisco
Juniper
Arista
H3C
Huawei

[root@CentOS-Python ~]#

#在a模式下使用write()函數修改文字內容
>>> f = open('test.txt','a')
>>> f.write('Avaya')
>>> f.close()

#文字修改後的內容
[root@CentOS-Python ~]# cat test.txt
```

```
Cisco
Juniper
Arista
H3C
Huawei
Avaya

[root@CentOS-Python ～]#

#在a+模式下使用write()函數修改文字內容
>>> f = open('test.txt','a+')
>>> f.write('Aruba')
>>> f.close()

#文字修改後的內容
[root@CentOS-Python ～]# cat test.txt
Cisco
Juniper
Arista
H3C
Huawei
Avaya
Aruba

[root@CentOS-Python ～]#
```

3.3.4 with 敘述

每次用 open() 函數開啟一個檔案，該檔案都將一直處於開啟狀態。這一點
我們可以用 closed 方法來驗證：如果檔案處於開啟狀態，則 closed 方法傳
回 False；如果檔案已被關閉，則 closed 方法傳回 True。

```
>>> f = open('test.txt')
>>> f.closed
False
```

```
>>> f.close()
>>> f.closed
True
>>>
```

這種每次都要手動關閉檔案的做法略顯麻煩,可以使用 with 敘述來管理檔案物件。用 with 敘述開啟的檔案將被自動關閉,舉例如下。

```
>>> with open('test.txt') as f:
...     print f.read()
...
Cisco
Juniper
Arista
H3C
Huawei

>>> f.closed
True
>>>
```

這裡沒有使用 close() 來關閉檔案,因為 with 敘述已經自動將檔案關閉(closed 方法的傳回值為 True)。

3.4 自訂函數

函數是已經組織好的可以被重複使用的一組程式區塊,它的作用是用來加強程式的重複使用率。在 Python 中,有很多內建函數(Built-in Function),例如前面已經講到的 type()、dir()、print()、int()、str()、list()、open() 等,在安裝好 Python 後就能立即使用。除了上述內建函數,我們也可以透過建立自訂函數(User-Defined Function)來完成一些需要重複使用的程式區塊,加強工作效率。

3.4.1　函數的建立和呼叫

在 Python 中，我們使用 def 敘述來自訂函數。def 敘述後面接函數名稱和括號 ()，括號裡根據情況可帶有參數也可沒有參數。在自訂函數建立好後，要將該函數呼叫才能獲得函數的輸出結果（即使該函數沒有參數），舉例如下。

```
#帶有參數的自訂函數
>>> def add(x，y):
...     result = x + y
...     print (result)
...
>>> add(1，2)
3
>>>

#沒有參數的自訂函數
>>> def name():
...     print ('Parry)'
...
>>>
>>> name()
Parry
>>>
```

不管自訂函數是否帶有參數，**函數都不能在建立前就被呼叫**，例如下面這段用來求一個數的二次方的指令稿。

```
[root@CentOS-Python ~]# cat test.txt
square(10)

def square(x):
 squared = x ** 2
 print (squared)
```

```
[root@CentOS-Python ~]# python3.8 test.txt
Traceback (most recent call last):
  File "test.txt", line 1, in <module>
    square(10)
NameError: name 'square' is not defined
[root@CentOS-Python ~]#
```

執行該指令稿後顯示出錯,原因就是在還沒有建立函數的情況下提前呼叫
了該函數,正確寫法如下。

```
[root@CentOS-Python ~]# cat test.txt

def square(x):
 squared = x ** 2
 print squared

square(10)

[root@CentOS-Python ~]# python test.txt
100
[root@CentOS-Python ~]#
```

3.4.2 函數值的傳回

任何函數都需要傳回一個值才有意義。自訂函數可以用 print 和 return 兩種
敘述來傳回一個值,**如果函數中沒有使用 print 或 return,則該函數只會傳
回一個空值(None)**,舉例如下。

```
>>> def add_1(x):
...    x = x+1
>>> print add_1(1)
None
>>>
```

❑ print 和 return 的區別

print 用來將傳回值列印輸出在控制端上,以便讓使用者看到,但是該傳回值不會被儲存下來。也就是說,如果將來把該函數設定值給一個變數,該變數的值將仍然為空值(None),舉例如下。

```
[root@CentOS-Python ~]# cat test.txt
def name():
    print ('Parry')


name()
a = name()
print (a)


[root@CentOS-Python ~]# python3.8 test.txt
Parry
Parry
None
[root@CentOS-Python ~]#
```

注意這裡傳回了兩個 Parry,是因為除了 name() 直接呼叫函數,將函數設定值給一個變數時,例如這裡的 a = name(),也會觸發呼叫函數的效果。

而 return 則恰恰相反,在呼叫函數後,return 的傳回值不會被列印輸出在控制端上,如果想看輸出的傳回值,則要在呼叫函數時在函數前加上 **print。但是傳回值會被儲存下來,如果把該函數設定值給一個變數,則該變數的值即該函數的傳回值**,舉例如下。

```
[root@CentOS-Python ~]# cat test.txt
def name():
    return 'Parry'


name()
a = name()
print (a)
```

```
[root@CentOS-Python ～]# python test.txt
Parry
[root@CentOS-Python ～]#
```

3.4.3 巢狀結構函數

函數支援巢狀結構,也就是一個函數可以在另一個函數中被呼叫,舉例如下。

```
[root@CentOS-Python ～]# cat test.txt

def square(x):
    result = x ** 2
    return result

def cube(x):
    result = square(x) * x
        return result

cube(3)
print (cube(3))

[root@CentOS-Python ～]# python test.txt
27
```

我們首先建立一個求 2 次方的函數 square(x)(注意這裡用的是 return,不是 print,否則傳回值將是 None,不能被其他函數套用)。然後建立一個求 3 次方的函數 cube(x),在 cube(x) 中我們套用了 square(x),並且也使用了 return 來傳回函數的結果。最後分別使用了 cube(3) 和 print(cube(3)) 來展示和證明 3.4.2 節中提到的「在呼叫函數後,return 的傳回值不會被列印輸出在控制端上,如果想看輸出的傳回值,則要在呼叫函數時在函數前面加上 print() 函數」,因為這裡只能看到一個輸出結果 27。

3.5 模組

在第 1 章已經講過，Python 的執行模式大致分為兩種：一種是使用解譯器
的互動模式，另一種是執行指令稿的指令稿模式。在互動模式下，一旦退
出解譯器，那麼之前定義的變數、函數及其他所有程式都會被清空，一切
都需要從頭開始。因此，如果想寫一段較長、需要重複使用的程式，則最
好使用編輯器將程式寫進指令稿，以便將來重複使用。

不過在網路工程師的日常工作中，隨著網路規模越來越大，需求越來越
多，對應的程式也將越寫越多。為了方便維護，我們可以把其中一些常用
的自訂函數分出來寫在一個獨立的指令檔中，然後在互動模式或指令稿模
式下將該指令檔匯入（import）以便重複使用，這種在互動模式下或其他
指令稿中被匯入的指令稿被稱為**模組（Module）**。在 Python 中，我們使用
import 敘述來匯入模組，**指令稿的檔案名稱（不包含副檔名 .py）即模組
名稱**。

被用作模組的指令稿中可能帶自訂函數，也可能不帶，下面將分別舉例
說明。

3.5.1 不帶自訂函數的模組

首先建立一個指令稿，將其命名為 script1.py，該指令稿的程式只有一行，
即列印內容「這是指令稿 1.」。

```
[root@CentOS-Python ~]# cat script1.py
# coding=utf-8
print ("這是指令稿1.")
[root@CentOS-Python ~]#
```

然後建立第二個指令稿，將其命名為 script2.py。在指令稿 2 裡，我們將
使用 import 敘述匯入指令稿 1（import script1），列印內容「這是指令稿

2.」，並執行指令稿 2。

```
[root@CentOS-Python ~]# cat script2.py
# coding=utf-8
import script1
print "這是指令稿2."

#執行script2
[root@CentOS-Python ~]# python3.8 script2.py
這是指令稿1.
這是指令稿2.
[root@CentOS-Python ~]#
```

可以看到，在執行指令稿 2 後，我們同時獲得了「這是指令稿 1.」和「這是指令稿 2.」的列印輸出內容，其中，「這是指令稿 1.」正是指令稿 2 透過 import script1 匯入指令稿 1 後獲得的。

3.5.2 帶自訂函數的模組

首先修改指令稿 1 的程式，建立一個 test() 函數，該函數的程式只有一行，即列印內容「這是帶函數的指令稿 1.」。

```
[root@CentOS-Python ~]# cat script1.py
# coding=utf-8
def test():
    print ("這是帶函數的指令稿1.")
[root@CentOS-Python ~]#
```

然後修改指令稿 2 的程式，呼叫指令稿 1 的 test() 函數，因為是指令稿 1 的函數，所以需要在函數名稱前加入模組名稱，即 script1.test()。

```
[root@CentOS-Python ~]# cat script2.py
# coding=utf-8
import script1
print "這是指令稿2."
```

```
script1.test()

#執行script2
[root@CentOS-Python ~]# python script2.py
這是指令稿2.
這是帶函數的指令稿1.
```

3.5.3　Python 內建模組和協力廠商模組

除了上述兩種使用者自己建立的模組，Python 還有內建模組及需要透過 pip 下載安裝的協力廠商模組（也叫作協力廠商函數庫），下面分別說明。

1. Python 內建模組

Python 有大量的內建模組直接透過 import 就可以使用，後面實驗部分的案例程式將重點說明這些內建模組的使用。這裡僅舉一例，我們可以使用 os 這個內建函數來發送 Ping 封包，判斷網路目標是否可達，指令稿程式及解釋如下。

```
import os
hostname = 'www.cisco.com'
response = os.system("ping -c 1 " + hostname)
if response == 0:
    print (hostname + ' is reachable.')
else:
    print (hostname + ' is not reachable.')
```

- os 是很常用的 Python 內建模組，os 是 operating system 的簡稱。顧名思義，它是用來與執行程式的主機作業系統互動的，後面實驗部分的案例程式會介紹關於 os 在網路運行維護中的許多用法。
- 這裡我們使用 os 模組的 system() 函數來直接在主機裡執行 ping -c 1 http://www.cisco.com 這行指令，也就是向 http://www.cisco.com 發送 1 個 Ping 封包。注意，我們做程式示範的主機是 CentOS，CentOS 中指

定 Ping 封包個數的參數為 -c。如果在 Windows 中執行這段指令稿,因為 Windows 中指定 Ping 封包個數的參數為 -n,則需要將程式裡的參數 -c 換成 -n,也就是 "ping -n 1 " + hostname !

■ os.system() 的傳回值為一個整數。如果傳回值為 0,則表示目標可達;如果為非 0,則表示無法連接。我們將 os.system() 的傳回值指定給變數 response,然後寫一個簡單的 if...else 判斷敘述來作判斷。如果 response 的值等於 0,則列印目標可達的資訊,反之則列印目標無法連接的資訊。

執行程式看效果。

```
[root@CentOS-Python ~]# python3.8 test.py
PING e2867.dsca.akamaiedge.net (104.86.224.155) 56(84) bytes of data.
64 bytes from a104-86-224-155.deploy.static.akamaitechnologies.com
(104.86.224.155): icmp_seq=1 ttl=128 time=115 ms

--- e2867.dsca.akamaiedge.net ping statistics ---
1 packets transmitted, 1 received, 0% packet loss, time 0ms
rtt min/avg/max/mdev = 115.206/115.206/115.206/0.000 ms
www.cisco.com is reachable.
[root@CentOS-Python ~]#
```

除了 os,能實現 ping 指令功能的 Python 內建模組還有很多,如 subprocess。當在 Python 中使用 os 和 subprocess 模組來執行 ping 指令時,兩者都有個「小缺陷」,就是它們都會顯示作業系統執行 ping 指令後的回應內容,如果想讓 Python「靜悄悄」地 ping,則可以使用協力廠商模組 pythonping,關於 pythonping 模組的用法在實驗部分中將提到。

2. Python 協力廠商模組

Python 協力廠商模組需要從 pip 下載安裝,pip 隨 Python 版本的不同有對應的 pip2、pip3 和 pip3.8,在 CentOS 8 裡已經內建了 pip2 和 pip3,我們在第 1 章介紹 Python 3.8 在 Windows 和 Linux 的安裝時已經安裝了 pip3.8,可以在 Windows 命令列及 CentOS 8 下輸入 pip3.8 來驗證,如下圖所示。

```
C:\Users\admin>
C:\Users\admin>pip3.8

Usage:
  pip3.8 <command> [options]

Commands:
  install                     Install packages.
  download                    Download packages.
  uninstall                   Uninstall packages.
  freeze                      Output installed packages in requirements format.
  list                        List installed packages.
  show                        Show information about installed packages.
  check                       Verify installed packages have compatible dependencies.
  config                      Manage local and global configuration.
  search                      Search PyPI for packages.
  wheel                       Build wheels from your requirements.
  hash                        Compute hashes of package archives.
  completion                  A helper command used for command completion.
  debug                       Show information useful for debugging.
  help                        Show help for commands.
```

```
root@CentOS-Python:~
[root@CentOS-Python ~]# pip3.8

Usage:
  pip3.8 <command> [options]

Commands:
  install                     Install packages.
  download                    Download packages.
  uninstall                   Uninstall packages.
  freeze                      Output installed packages in requirements format.
  list                        List installed packages.
  show                        Show information about installed packages.
  check                       Verify installed packages have compatible dependen
cies.
  config                      Manage local and global configuration.
  search                      Search PyPI for packages.
  wheel                       Build wheels from your requirements.
  hash                        Compute hashes of package archives.
  completion                  A helper command used for command completion.
  debug                       Show information useful for debugging.
  help                        Show help for commands.
```

對網路工程師來説，最常用的 Python 協力廠商模組無疑是用來 SSH 登入
網路裝置的 Paramiko 和 Netmiko。首先使用指令 pip3.8 install Paramiko 和
pip3.8 install Netmiko 來分別安裝它們（如果你的 Python 不是 3.8.x 版本，
而是其他 Python 3 版本，則將 pip3.8 取代成 pip3 即可）。在安裝之前，請
確認你的 CentOS 8 主機或虛擬機器能夠連上外網。

```
[root@CentOS-Python ~]#pip3.8 install Paramiko
[root@CentOS-Python ~]#pip3.8 install Netmiko
```

註：如果你使用的是 Python 2，則需要使用 pip2 install Netmiko == 2.4.2 來安裝 Netmiko，因為 2020 年 1 月後，所有透過 pip 安裝的 Netmiko 都預設只支援 Python 3。

使用 pip 安裝好 Paramiko 和 Netmiko 後，開啟 Python 測試是否可以使用 import Paramiko 和 import Netmiko 來參考它，如果沒顯示出錯，則說明安裝成功。

```
>>> import Paramiko
>>> import Netmiko
Traceback (most recent call last):
  File "<stdin>", line 1, in <module>
ModuleNotFoundError: No module named 'Netmiko'
>>>
```

由上面程式可以看到，import Paramiko 後沒有任何問題，但是 import Netmiko 卻收到解譯器顯示出錯 "ModuleNotFoundError: No module named 'Netmiko'"，這是因為為了效果示範，筆者故意只安裝了 Paramiko 而沒有安裝 Netmiko。關於 Paramiko 模組的實際用法，後面將詳細介紹，這裡僅做示範來說明如何在 Python 中透過 pip 安裝協力廠商模組。

3.5.4 from ... import ...

對於帶自訂函數的模組，除了 import 敘述，我們還可以使用 from... import... 來匯入模組。它的作用是省去了在呼叫模組函數時必須加上模組名稱的麻煩，該敘述實際的格式為 from [模組名稱] import [函數名稱]，舉例如下。

將指令稿 2 的程式修改如下，然後執行指令稿 2。

```
[root@CentOS-Python ~]# cat script2.py
# coding=utf-8
from script1 import test
```

```
print ("這是指令稿2.")
test()

#執行script2
[root@CentOS-Python ~]# python script2.py
這是指令稿2.
這是帶函數的指令稿1.
```

因為我們在指令稿 2 中使用了 from script1 import test，所以在呼叫指令稿
1 的 test() 函數時不再需要寫成 script1.test()，直接用 test() 即可。

3.5.5 if __name__ == '__main__':

在 3.5.1 節所舉的實例中，指令稿 2 在匯入了模組指令稿 1 後，立即就參
考了指令稿 1 的程式，即 print (" 這是指令稿 1.")。有時我們只是希望使用
所匯入模組中的部分函數，並不希望一次性全部引用模組中的程式內容，
這時就可以用 if __name__ == '__main__': 這個判斷敘述來達到目的。

在基本語法中已經講過，Python 中前後帶雙底線的變數叫作內建變數，如
__name__，關於內建變數的內容已經超出了本書的範圍。我們可以這麼來
了解：**如果一個 Python 指令檔中用到了 if __name__ == '__main__': 判
斷敘述，則所有寫在其下面的程式都將不會在該指令稿被其他指令稿用作
模組匯入時被執行。**

我們首先將指令稿 1 和指令稿 2 的程式分別修改如下，在指令稿 1 中將
print (" 這是指令稿 1.") 寫在 if __name__ == '__main__': 的下面。

```
[root@CentOS-Python ~]# cat script1.py
# coding=utf-8
if __name__ == '__main__':
    print ("這是指令稿1.")
[root@CentOS-Python ~]#
```

```
[root@CentOS-Python ~]# cat script2.py
# coding=utf-8
import script1
print ("這是指令稿2.")
[root@CentOS-Python ~]#
```

然後執行指令稿 2，發現雖然指令稿 2 匯入了指令稿 1，但是輸出結果中卻不再出現「這是指令稿 1.」的輸出結果。

```
[root@CentOS-Python ~]# python3.8 script2.py
這是指令稿2.
[root@CentOS-Python ~]#
```

3.6 正規表示法

在網路工程師的日常工作中，少不了要在路由器、交換機、防火牆等裝置的命令列中使用各種 show 或 display 指令來查詢設定、裝置資訊或進行校正，例如思科和 Arista 裝置上最常見的 show run、show log、show interface 和 show ip int brief 等指令。通常這些指令輸出的資訊和回應內容過多，需要用管線符號 |（Pipeline）配合 grep（Juniper 裝置）或 include/exclude/begin（思科裝置）等指令來比對或篩選我們所需的資訊。舉個實例，要查詢一台 24 埠的思科 2960 交換機目前有多少個通訊埠是 Up 的，可以使用指令 show ip int brief | i up，程式如下。

```
2960#show ip int b | i up
FastEthernet0/2 unassigned YES unset up up
FastEthernet0/10  unassigned YES unset up up
FastEthernet0/11  unassigned YES unset up up
FastEthernet0/12  unassigned YES unset up up
FastEthernet0/15  unassigned YES unset up up
FastEthernet0/23  unassigned YES unset up up
```

```
FastEthernet0/24  unassigned YES unset up up
GigabitEthernet0/2  unassigned YES unset up up
Loopback0 unassigned YES NVRAM up up
```

管線符號後面的指令部分（i up）即本節將要講的**正規表示法（Regular Expression**）。另外，如果你是 CCIE 或是擁有數年從業經驗、負責過大型企業網或電信業者網路運行維護的資深網路工程師，那麼應該不會對 BGP（Border Gateway Protocol，邊界閘道協定）中出現的自治域路徑存取控制串列（as-path access-list）感到陌生。as-path access-list 是正規表示法在電腦網路中最典型的應用，例如下面的 as-path access-list 1 中的 ^4$ 就是一個標準的正規表示法。

```
ip as-path access-list 1 permit ^4$
```

3.6.1 什麼是正規表示法

根據維基百科的解釋：

正規表示法（Regular Expression，在程式中常簡寫為 regex、regexp 或 RE），又稱**正規表示式、正規標記法、正規運算式、正則運算式、正常標記法**，是電腦科學的概念。正規表示法使用單一字串來描述、比對一系列比對某個句法規則的字串。在很多文字編輯器中，正規表示法通常被用來檢索、取代那些比對某個模式的文字。

在 Python 中，我們使用正規表示法來對文字內容做解析（Parse），即從一組指定的字串中透過指定的字元來搜尋和比對我們需要的「模式」（**Pattern，可以把「模式」了解成我們要搜尋和符合的「關鍵字」**）。正規表示法本身並不是 Python 的一部分，它擁有自己獨特的語法及獨立的處理引擎，效率上可能不如字串附帶的各種方法，但它的功能遠比字串附帶的方法強大得多，因為它同時支援**精確比對（Exact Match**）和**模糊比對**（Wildcard Match）。

3.6.2 正規表示法的驗證

怎麼知道自己的正規表示法是否寫正確了呢？方法很簡單，可以透過線上正規表示法模擬器來驗證自己寫的正規表示法是否正確。線上正規表示法模擬器很多，透過搜尋引擎很容易找到。線上正規表示法模擬器的使用也很簡單，通常只需要提供文字內容（即字串內容），以及用來比對模式的正規表示法即可。下圖的 regex101 就是筆者常用的線上正規表示法模擬器，後面在舉例說明正規表示法時，筆者會提供在該模擬器上驗證的畫面來佐證。

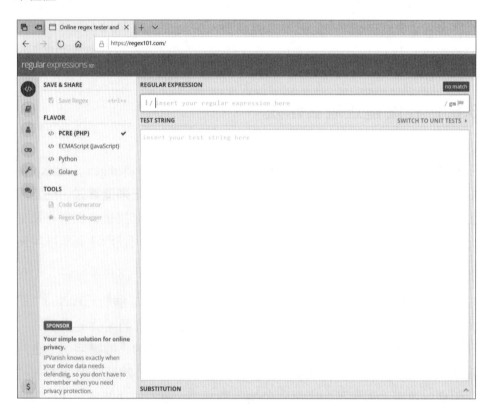

.

3.6.3 正規表示法的規則

正規表示法是一套十分強大但也十分複雜的字元比對工具，本節只選取其中部分網路運行維護中常用並且適合網路工程師學習的基礎知識，然後配合案例說明。首先來了解正規表示法的**精確比對**和**模糊比對**，其中模糊比對又包含**比對符號**（Matching Characters）和**特殊序列**（Special Sequence）。

1. 精確比對

精確比對即明文列出我們想要符合的模式。例如上面講到的在 24 埠的思科 2960 交換機裡尋找 Up 的通訊埠，我們就在管線符號 | 後面明文列出模式 Up。又例如我們想在下面的交換機記錄檔中找出所有記錄檔類型為 %LINK-3-UPDOWN 的記錄檔，那我們按照要求明文列出模式 %LINK-3-UPDOWN 即可，即指令 show logging | i %LINK-3- UPDOWN。

```
2960#show logging | i %LINK-3-UPDOWN
000458: Feb 17 17:10:31.906: %LINK-3-UPDOWN: Interface FastEthernet0/2，
changed state to down
```

但是如果同時有多種記錄檔類型需要比對，程式如下。

```
000459: Feb 17 17:10:35.202: %LINK-3-UPDOWN: Interface FastEthernet0/2，
changed state to up
000460: Feb 17 17:10:36.209: %LINEPROTO-5-UPDOWN: Line protocol on Interface
FastEthernet0/2，changed state to up
000461: Feb 17 22:39:26.464: %SSH-5-SSH2_SESSION: SSH2 Session request
from 10.1.1.1 (tty = 0) using crypto cipher 'aes128-cbc'，hmac 'hmac-sha1'
Succeeded
000462: Feb 17 22:39:27.748: %SSH-5-SSH2_USERAUTH: User 'test' authentication
for SSH2 Session from 10.1.1.1 (tty = 0) using crypto cipher 'aes128-cbc'，
hmac 'hmac-sha1' Succeeded
```

這時，再用精確比對就顯得很笨拙，因為要透過指令 show logging | i %LINK-3- UPDOWN|%LINEPROTO-5-UPDOWN|%SSH-5-SSH2_SESSION|%SSH-5-SSH2_USERAUTH 把所有有興趣的記錄檔類型都明文列出來，這裡只有 4 種記錄檔類型還比較容易，如果有幾十上百種記錄檔類型去比對，再進行精確符合的工作量就會很大，這時我們需要借助模糊比對來完成這項任務。

2. 模糊比對

模糊比對包含比對符號和特殊序列，下面分別說明。

正規表示法中常見的比對符號如下表所示。

比對符號	用　法
.	比對除分行符號外的所有字元，比對次數為 1 次
*	用來比對緊靠該符號左邊的符號，比對次數為 0 次或多次
+	用來比對緊靠該符號左邊的符號，比對次數為 1 次或多次
?	用來比對緊靠該符號左邊的符號，比對次數為 0 次或 1 次
{m}	用來比對緊靠該符號左邊的符號，指定比對次數為 m 次。例如字串 'abbbbcccd'，使用 ab{2} 將比對到 abb，使用 bc{3}d 將比對到 bcccd
{m,n}	用來比對緊靠該符號左邊的符號，指定比對次數為最少 m 次，最多 n 次。舉例來說，如果字串為 'abbbbcccd'，使用 ab{2,3} 將只能比對到 abb；如果字串為 'abbbcccdabbccd'，使用 ab{2,3} 將能同時比對到 abbb 和 abb；如果字串內容為 'abcd'，使用 ab{2,3} 將比對不到任何東西
{m,}	用來比對緊靠該符號左邊的符號，指定比對次數最少為 m 次，最多無限次
{,n}	用來比對緊靠該符號左邊的符號，指定比對次數最少為 0 次，最多為 n 次
\	逸出字元，用來比對上述「比對符號」。例如字串內容中出現了問號 ? ，而又想精確比對這個問號，那就要用 \ ? 來進行比對。除此之外，\ 也用來表示一個特殊序列，特殊序列將在下節講到

比對符號	用　法
[]	表示字元集合，用來精確比對。例如要精確比對一個數字，可以使用 [0-9]；如果要精確比對一個小寫字母，可以用 [a-z]；如果要精確比對一個大寫字母，可以用 [A-Z]；如果要比對一個數字、字母或底線，可以用 [0-9a-zA-Z_]。另外，在 [] 中加 ^ 表示取非，例如 [^0-9] 表示比對一個非數字的字元，[^a-z] 表示比對一個非字母的字元，依此類推
|	表示「或比對」（兩項中比對其中任意一項），例如要比對 FastEthernet 和 GigabitEthernet 這兩種通訊埠名稱，可以寫作 Fa|Gi
(⋯)	組合，比對括號內的任意正規表示法，並表示組合的開始和結尾，例如 (b|cd)ef 表示比對 bef 或 cdef

3. 貪婪比對

*、+、?、{m}、{m,} 和 {m,n} 這 6 種比對符號預設都是貪婪符合的，即會盡可能多地去比對符號合條件的內容。

假設指定的一組字串為 "xxzyxzyz"，我們使用正規表示法模式 x.*y 來做比對（註：**精確比對和模糊比對可以混用**）。在比對到第一個 "x" 後，開始比對 .*，因為 . 和 * 預設是貪婪符合的，它會一直往後比對，直到比對到最後一個 "y"，因此比對結果為 "xxzyxzy"，可以在 regex101 上驗證，如下圖所示。

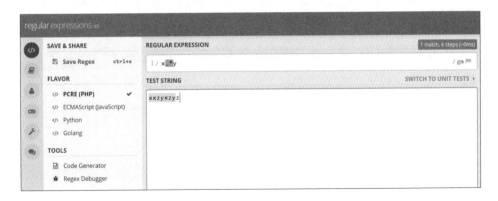

3-47

又假設指定的字串依然為 "xxzyxzyz"，我們使用正規表示法模式 xz.*y 來做比對，在比對到第一個 "xz" 後，開始比對「貪婪」的 .*，這裡將一直往後比對，直到最後一個 "y"，因此比對結果為 "xzyxzy"，在 regex101 上驗證，如下圖所示。

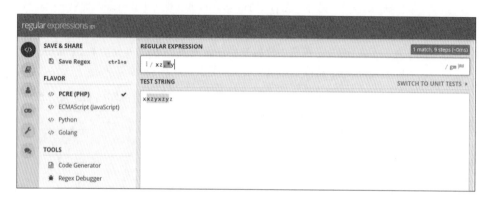

4. 非貪婪比對

要實現非貪婪比對很簡單，就是在上述 6 種貪婪比對符號後面加上問號 ? 即可，即 *?、+?、??、{m}?、{m,}? 和 {m,n}?。

假設指定的另一組字串為 "xxzyzyz"（注意不是之前的 "xxzyxzyz"），我們使用正規表示法模式 x.*?y 來做比對。在比對到第一個 "x" 後，開始比對 .*?，因為 .*? 是非貪婪符合的，它在比對到第一個 "y" 後便隨即停止，因此比對結果為 "xxzy"，在 regex101 上驗證，如下圖所示。

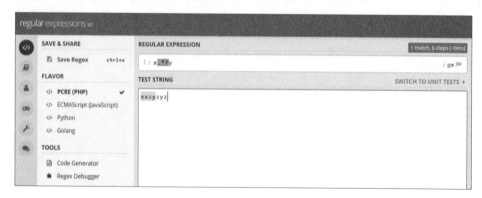

又假設指定的字串依然為 "xxzyzyz"，我們使用正規表示法模式 xz.*?y 來
做比對，在比對到第一個 "xz" 後，開始比對 .*?，它在比對到第一個 "y"
後便隨即停止，因此比對結果為 "xzy"，在 regex101 上驗證，如下圖所示。

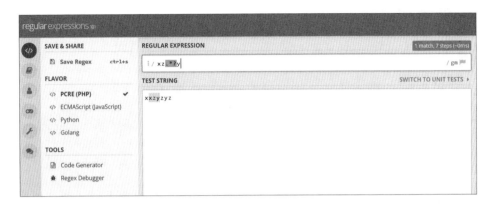

❑ 正規表示法中常見的特殊序列

特殊序列由逸出符號 \ 和一個字元組成，常見的特殊序列及其用法如下表
所示。

特殊序列	用　法
\d	比對任意一個十進位數字，相等於 [0-9]
\D	\d 取非，比對任意一個非十進位數字，相等於 [^0-9]
\w	比對任意一個字母、十進位數字及底線，相等於 [a-zA-Z0-9]
\W	\w 取非，相等於 [^a-zA-Z0-9]
\s	比對任意一個空白字元，包含空格、分行符號 \n
\S	\s 取非，比對任意一個不可為空白字元

模糊比對在正規表示法中很常用，前面精確比對中提到的比對思科交換機
記錄檔類型的實例可以用模糊比對來處理，例如我們要在下面的記錄檔
中同時比對 %LINK-3- UPDOWN、%LINEPROTO-5-UPDOWN、%SSH-5-

SSH2_SESSION 和 %SSH-5-SSH2_USE RAUTH 4 種記錄檔類型，用正規表示法 %\w{3,9}-\d-\w{6,13} 即可完全符合。

```
000459: Feb 17 17:10:35.202: %LINK-3-UPDOWN: Interface FastEthernet0/2，
changed state to up
000460: Feb 17 17:10:36.209: %LINEPROTO-5-UPDOWN: Line protocol on Interface
FastEthernet0/2，changed state to up
000461: Feb 17 22:39:26.464: %SSH-5-SSH2_SESSION: SSH2 Session request
from 10.1.1.1 (tty = 0) using crypto cipher 'aes128-cbc'，hmac 'hmac-sha1'
Succeeded
000462: Feb 17 22:39:27.748: %SSH-5-SSH2_USERAUTH: User 'test' authentication
for SSH2 Session from 10.1.1.1 (tty = 0) using crypto cipher 'aes128-cbc'，
hmac 'hmac-sha1' Succeeded
```

關於該正規表示法 %\w{3,9}-\d-\w{4,13} 是如何完整比對上述 4 種記錄檔類型的說明如下。

首先將 %\w{3,9}-\d-\w{4,13} 拆分為 6 部分。

第 1 部分：% 用來精確比對百分比 "%"（4 種記錄檔全部以 "%" 開頭），如下圖所示。

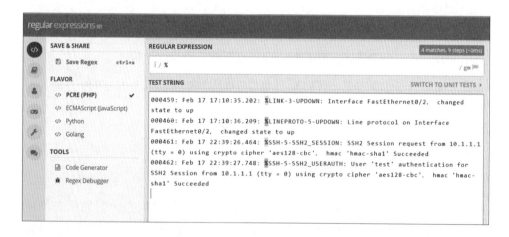

第 2 部分：\w{3,9} 用來比對 "SSH" 和 "LINEPROTO"，如下圖所示。

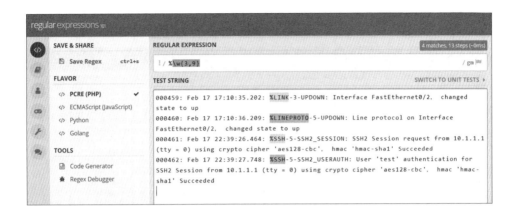

第 3 部分：- 用來精確比對第一個 "-"，如下圖所示。

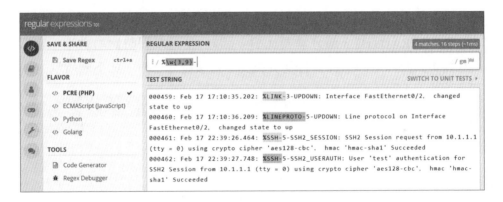

第 4部分：\d 用來比對數字 "3" 或 "5"，如下圖所示。

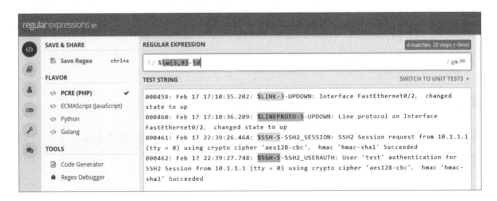

第 5 部分：- 用來精確比對第二個 "-"，如下圖所示。

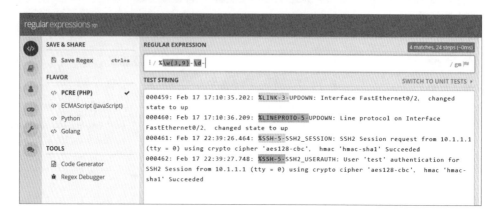

第 6 部分：由上圖可知，至此還剩下 "UPDOWN"、"SSH2_SESSION"、"SSH2_USERAUTH" 3 部分需要比對，因為 \w 可以同時比對數字、字母及底線，因此，用 \w{6,13} 即可完整比對最後這 3 部分，如下圖所示。

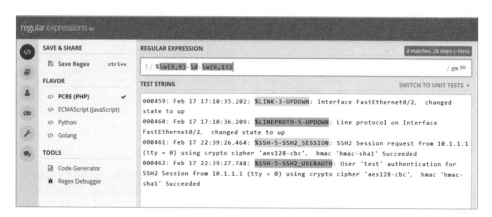

3.6.4 正規表示法在 Python 中的應用

在 Python 中，我們使用 import re 來匯入正規表示法這個內建模組（無須使用 pip 來安裝）。

```
import re
```

首先使用 dir() 函數來看下 re 模組中有哪些內建函數和方法。

```
>>> import re
>>> dir(re)
['DEBUG', 'DOTALL', 'I', 'IGNORECASE', 'L', 'LOCALE', 'M', 'MULTILINE',
'S', 'Scanner', 'T', 'TEMPLATE', 'U', 'UNICODE', 'VERBOSE', 'X',
'_MAXCACHE', '__all__', '__builtins__', '__doc__', '__file__',
'__name__', '__package__', '__version__', '_alphanum', '_cache',
'_cache_repl', '_compile', '_compile_repl', '_expand', '_locale',
'_pattern_type', '_pickle', '_subx', 'compile', 'copy_reg', 'error',
'escape', 'findall', 'finditer', 'match', 'purge', 'search', 'split',
'sre_compile', 'sre_parse', 'sub', 'subn', 'sys', 'template']
```

其中，網路工程師較常用的 Python 正規表示法的函數主要有 4 種，分別為 re.match()、re.search()、re.findall() 和 re.sub()，下面對它們分別說明。

1. re.match()

re.match() 函數用來在字串的起始位置比對指定的模式，如果比對成功，則 re.match() 的傳回值為比對到的物件。**如果想檢視比對到的物件的實際值，則還要對該物件呼叫 group() 函數。**如果比對到的模式不在字串的起始位置，則 re.match() 將傳回空值（None）。

re.match() 函數的語法如下。

```
re.match(pattern, string, flags=0)
```

pattern 即我們要符合的正規表示法模式。string 為要符合的字串。flags 為標示位，用來控制正規表示法的比對方式，如是否區分大小寫、是否多行比對等。flags 為可選項，不是很常用。

舉例如下。

```
[root@CentOS-Python ~]# python3.8
Python 3.8.2 (default, Apr 27 2020, 23:06:10)
[GCC 8.3.1 20190507 (Red Hat 8.3.1-4)] on linux
Type "help", "copyright", "credits" or "license" for more information.
>>> import re
>>> test = 'Test match() function of regular expression.'
>>> a = re.match(r'Test', test)
>>> print (a)
<re.Match object; span=(0, 4), match='Test'>
>>> print (a.group())
Test
>>>
```

我 們 使 用 re.match() 函 數， 從 字 串 "Test match() function of regular expression." 裡精確比對模式 "Test"，因為 "Test" 位於該段字串的起始位置，所以比對成功，並且傳回一個比對到的物件 <re.Match object; span=(0, 4), match='Test'>（即用 print(a) 看到的內容），為了檢視該物件的實際值，我們可以對該物件呼叫 group() 方法，獲得實際值 "Test"（即用 print (a.group()) 看到的內容），**該值的資料類型為字串**，group() 函數在 Python 的正規表示法中很常用，務必熟練使用。

如果我們不從字串的起始位置去比對，而是去比對中間或尾端的字串內容，則 re.match() 將比對不到任何內容，進一步傳回空值（None）。 例如我們嘗試比對 "function" 這個詞，因為 "function" 不在 "Test match() function of regular expression." 的開頭，所以 re.match() 的傳回值為 None。

```
>>> import re
>>> test = 'Test match() function of regular expression.'
>>> a = re.match(r'function'，test)
>>> print (a)
None
>>>
```

在上面兩個實例中,我們分別在模式 "Test" 和 "function" 的前面加上了一個 r,這個 r 代表原始字串(Raw String)。在 Python 中,原始字串主要用來處理特殊字元所產生的問題,例如前面講到的逸出字元 \ 就是一種特殊字元,它會產生很多不必要的問題。舉個實例,假設你要在 Windows 中用 Python 的 open() 函數開啟一個檔案,程式如下。

```
>>> f = open('C:\Program Files\test.txt', 'r')
IOError: [Errno 2] No such file or directory: 'C:\\Program Files\test.txt'
>>>
```

這時,你會發現該檔案打不開了,Python 傳回了一個 "IOError: [Errno 2] No such file or directory: 'C:\\Program Files\test.txt'" 錯誤,這是因為 "\t" 被當作了不屬於檔案名稱的特殊符號。解決的辦法也很簡單,就是在代表檔案所在路徑的字串前面使用原始字串。

```
>>> f = open(r'C:\Program Files\test.txt', 'r')
>>> print f
<open file 'C:\\Program Files\\test.txt', mode 'r' at 0x0000000002A01150>
>>>
```

在正規表示法中使用原始字串也是同樣的道理,只是還有更多其他原因,對這些原因的解釋已經超出了本書的範圍。網路工程師只需記住:**在正規表示法中,建議使用原始字串。**

2. re.search()

re.search() 函數和 re.match() 一樣,傳回值為字串,但是它比 re.match() 更靈活,因為它允許在字串的任意位置比對指定的模式。

re.search() 函數的語法如下。

```
re.search(pattern, string, flags=0)
```

前面我們用 re.match() 在 "Test match() function of regular expression." 中嘗試比對 "function" 不成功，因為 "function" 不在該字串的起始位置。我們改用 re.search() 來比對。

```
>>>import re
>>> test ='Test search() function of regular expression.'
>>> a = re.search(r'function', test)
>>> print (a)
<re.Match object; span=(14, 22), match='function'>
>>> print (a.group())
function
>>>
```

雖然 re.search() 可以在字串的任意位置比對模式，但是它和 re.match() 一樣一次只能比對一個字串內容，例如下面是某台路由器上 show ip int brief 指令的回應內容，我們希望用正規表示法來比對該回應內容中出現的所有 IPv4 位址。

```
Router#show ip int b
Interface IP-Address OK? Method Status Protocol
GigabitEthernet1/1 192.168.121.181 YES NVRAM up up
GigabitEthernet1/2 192.168.110.2  YES NVRAM up up
GigabitEthernet2/1 10.254.254.1 YES NVRAM up up
GigabitEthernet2/2 10.254.254.5 YES NVRAM up up
```

我們嘗試用 re.search() 來比對。

```
>>> test = '''Router#show ip int b
Interface IP-Address OK? Method Status Protocol
GigabitEthernet1/1 192.168.121.181 YES NVRAM up up
GigabitEthernet1/2 192.168.110.2  YES NVRAM up up
GigabitEthernet2/1 10.254.254.1 YES NVRAM up up
GigabitEthernet2/2 10.254.254.5 YES NVRAM up up '''
>>> ip_address = re.search(r'\d{1,3}\.\d{1,3}\.\d{1,3}\.\d{1,3}',test)
>>> print (ip_address.group())
192.168.121.181
```

我們用正規表示法 \d{1，3}\.\d{1，3}\.\d{1，3}\.\d{1，3} 作為模式來比對任意 IPv4 位址，注意我們在分割每段 IP 位址的 "." 前面都加了逸出符號 \，如果不加 \，寫成 \d{1，3}.\d{1，3}.\d{1，3}.\d{1，3}，則將比對到 GigabitEthernet1/1 中的 "1/1"，原因請讀者自行思考。

print (a.group()) 後 可 以 看 到 只 比 對 到 了 192.168.121.181 這 一 個 IPv4 位址，這就是 re.search() 的缺陷。如果想比對其他所有 IPv4 位址（192.168.110.2，10.254.254.1，10.254.254.5），則必須用下面要講的 re.findall()。

3. re.findall()

如果字串中有多個能被模式比對到的關鍵字，並且我們希望把它們全部比對出來，則要使用 re.findall()。同 re.match() 和 re.search() 不一樣，**re.findall() 的傳回值為串列，每個被模式比對到的字串內容分別是該串列中的元素之一。**

re.findall() 函數的語法如下。

```
re.findall(pattern, string, flags=0)
```

還是以上面嘗試比對 show ip int brief 指令的回應內容中所有 IPv4 位址的實例為例。

```
>>> test = '''Router#show ip int b
Interface IP-Address OK? Method Status Protocol
GigabitEthernet1/1 192.168.121.181 YES NVRAM up up
GigabitEthernet1/2 192.168.110.2  YES NVRAM up up
GigabitEthernet2/1 10.254.254.1 YES NVRAM up up
GigabitEthernet2/2 10.254.254.5 YES NVRAM up up '''
>>> ip_address = re.findall(r'\d{1,3}\.\d{1,3}\.\d{1,3}\.\d{1,3}',test)
>>> type(ip_address)
<type 'list'>
>>> print (ip_address)
```

```
['192.168.121.181', '192.168.110.2', '10.254.254.1', '10.254.254.5']
>>>
```

這裡成功透過 re.findall() 比對到所有的 4 個 IPv4 位址,每個 IPv4 位址分別為 re.findall() 傳回的串列中的元素。

4. re.sub()

最後要講的 re.sub() 函數是用來取代字串裡被比對到的字串內容,類似 Word 的取代功能。除此之外,它還可以定義最大取代數(Maximum Number of Replacement)來指定 sub() 函數所能取代的字串內容的數量,預設狀態下為全部取代。re.sub() 的傳回值是字串。

re.sub() 函數的語法如下。

```
a = re.match(pattern, replacement, string, optional flags)
```

re.sub() 函數的語法與前面講到的 re.match()、re.search()、re.findall() 3 個函數略有不同,re.sub() 函數裡多了一個 replacement 參數,它表示被取代後的字串內容。optional flags 可以用來指定所取代的字串內容的數量,如果只想取代其中 1 個字串內容,則可以將 optional flags 位設為 1;如果想取代其中的前兩個字串內容,則設為 2;依此類推。如果 optional flags 位元空缺,則預設狀態下為全部取代。

下面以某台路由器上的 ARP 表的輸出內容為例,我們用 re.sub() 來將其中所有的 MAC 位址全部取代為 1234.56ab.cdef。

```
>>> test = '''
... Router#show ip arp
... Protocol Address Age (min) Hardware Addr  Type  Interface
... Internet 10.1.21.1    -    b4a9.5aff.c845 ARPA  TenGigabitEthernet2/1
... Internet 10.11.22.1   51   b4a9.5a35.aa84 ARPA  TenGigabitEthernet2/2
... Internet 10.201.13.17 -    b4a9.5abe.4345 ARPA  TenGigabitEthernet2/3'''
>>> a = re.sub(r'\w{4}\.\w{4}\.\w{4}', '1234.56ab.cdef', test)
```

```
>>> print (a)

Router#show ip arp
Protocol   Address         Age (min)   Hardware Addr    Type    Interface
Internet   10.1.21.1          -         1234.56ab.cdef   ARPA    TenGigabitEthernet2/1
Internet   10.11.22.1         51        1234.56ab.cdef   ARPA    TenGigabitEthernet2/2
Internet   10.201.13.17       -         1234.56ab.cdef   ARPA    TenGigabitEthernet2/3
```

因為 optional flags 位空缺，所以預設將 3 個 MAC 位址全部取代成
1234.56ab.cdef。如果將 optional flags 位設為 1，則只會取代第一個 MAC
位址，效果如下。

```
>>> a = re.sub(r'\w{4}\.\w{4}\.\w{4}','1234.56ab.cdef','test',1)
>>> print (a)

Router#show ip arp
Protocol   Address         Age (min)   Hardware Addr    Type    Interface
Internet   10.1.21.1          -         1234.56ab.cdef   ARPA    TenGigabitEthernet2/1
Internet   10.11.22.1         51        b4a9.5a35.aa84   ARPA    TenGigabitEthernet2/2
Internet   10.201.13.17       -         b4a9.5abe.4345   ARPA    TenGigabitEthernet2/3
```

如果只希望取代某一個特定的 MAC 要怎麼做呢？方法也很簡單，這個問
題留給讀者朋友自行思考和實驗。

3.7 異常處理

異常處理（Exception Handling）是 Python 中很常用的基礎知識。通常在
寫完程式第一次執行指令稿時，我們難免會遇到一些程式錯誤。Python 中
有兩種程式錯誤：**語法錯誤**（Syntax Errors）**和異常**（Exceptions）。例如
下面這種忘了在 if 敘述尾端加冒號：就是一種典型的語法錯誤，Python 會
回覆一個 "SyntaxError: invalid syntax" 的顯示出錯資訊。

```
>>> if True
  File "<stdin>", line 1, in ?
    if True
          ^
SyntaxError: invalid syntax
>>>
```

有時一行敘述在語法上是正確的，但是執行程式後依然會引發錯誤，這種錯誤叫作異常。異常的種類很多，例如把零當作除數的「除零錯誤」（ZeroDivisionError）、變數還沒建立就被呼叫的「命名錯誤」（NameError）、資料類型使用有誤的「類型錯誤」（TypeError），以及嘗試開啟不存在的檔案時會遇到的「I/O 錯誤」（IOError）等都是很常見的異常。這些都是 Python 中常見的內建異常，也就是在沒有匯入協力廠商模組的情況下會遇到的異常，舉例如下。

```
#除零錯誤
>>> 100 / 0
Traceback (most recent call last):
  File "<stdin>", line 1, in <module>
ZeroDivisionError: division by zero
>>>

#命名錯誤
>>> print name
Traceback (most recent call last):
  File "<stdin>", line 1, in ?
NameError: name 'name' is not defined
>>>

#類型錯誤
>>> a = 10
>>> print ("There are " + a + "books.")
Traceback (most recent call last):
  File "<stdin>", line 1, in <module>
```

```
TypeError: can only concatenate str (not "int") to str
>>>


#I/O錯誤
>>> f = open('abc.txt')
Traceback (most recent call last):
  File "<stdin>", line 1, in <module>
FileNotFoundError: [Errno 2] No such file or directory: 'abc.txt'
>>>
```

除了這些常見的 Python 內建異常，從協力廠商匯入的模組也有自己獨有的
異常，例如後面實驗部分將重點講的 Paramiko 就有與 SSH 使用者名稱、
密碼錯誤相關的「AuthenticationException 異常」，以及網路裝置 IP 位址無
法連接導致的 Socket 模組的「socket.error 異常」。關於這些協力廠商模組
的使用及它們的異常處理在第 4 章的實驗 3 將重點介紹。

使用異常處理能加強程式的堅固性，幫助程式設計師快速修復程式中出現
的錯誤。在 Python 中，我們使用 try...except... 敘述來做異常處理，舉例如
下。

```
>>> for i in range(1，6):
...     print (i/0)
...
Traceback (most recent call last):
  File "<stdin>"，line 2，in <module>
ZeroDivisionError: integer division or modulo by zero
>>>

>>> for i in range(1，6):
...     try:
...         print (i/0)
...     except ZeroDivisionError:
...         print "Division by 0 is not allowed"
...
```

```
Division by 0 is not allowed
Division by 0 is not allowed
Division by 0 is not allowed
Division by 0 is not allowed
Division by 0 is not allowed
>>>
```

我們故意嘗試觸發除零錯誤，在沒有做異常處理時，如果 Python 解譯器發現 range(1，6) 傳回的第一個數字 1 試圖去和 0 相除，會馬上傳回一個 ZeroDivisonError，並且程式就此終止。在使用 try...except... 做異常處理時，我們使用 except 去主動捕捉除零錯誤這個異常類型（except ZeroDivisonError:），並告訴 Python 解譯器：在遇到除零錯誤時，**不要馬上終止程式**，而是列印出 "Division by 0 is not allowed" 這個資訊，告知使用者程式實際出了什麼問題，**然後繼續執行剩下的程式，直到完成**。正因如此，在遇到除零錯誤後，程式並沒有終止，而是接連列印出了 5 個 "Division by 0 is not allowed"（因為 range(1,6) 傳回了 1、2、3、4、5 共 5 個整數）。

為了更清楚地展示 try...except... 帶來的當執行程式遇到語法錯誤和異常時，Python 解譯器「不會馬上終止程式」的好處，我們把上述程式稍作修改。

```
>>> for i in range(1，6):
...     if i == 1:
...         print (i/0)
...     else:
...         print (i/1)
...
Traceback (most recent call last):
 File "<stdin>"，line 3，in <module>
ZeroDivisionError: integer division or modulo by zero
>>>
```

```
>>> for i in range (1,6):
...     try:
...         if i == 1:
...             print (i/0)
...         else:
...             print (i/1)
...     except ZeroDivisionError:
...         print 'Division by 0 is not allowed'
...
Division by 0 is not allowed
2
3
4
5
>>>
```

我們在原先的 for 循環敘述裡加入了 if 判斷敘述，只有當 range(1，6) 建立的整數 1 ～ 5 的數字為 1 時，才會觸發除零錯誤。當整數串列裡的數字不為 1 時，我們把它和 1 相除並列印出結果。在沒有使用異常處理的情況下，Python 解譯器在傳回 ZeroDivisionError: integer division or modulo by zero 後馬上終止了程式。使用異常處理後，在觸發除零錯誤時，程式並沒有被中斷，而是列印 "Division by 0 is not allowed" 後繼續執行。

另外，在上面兩個實例中，我們都在 except 敘述後面加入了 ZeroDivision Error，這種提前捕捉異常類型的做法是因為我們知道自己的程式將觸發除零錯誤，如果這時程式裡出現了另外一種異常，則程式還是會被中斷，因為 except 並沒有捕捉該異常。一般來說，我們很難記住 Python 中所有可能出現的異常類型，如果程式複雜，則無法預測指令稿中將出現什麼樣的異常類型，當出現這種情況時，我們要怎麼做異常處理呢？有兩種方法：一是在 except 敘述後面不接任何異常類型，直接寫成 "except:"。二是透過 Exceptions 捕捉所有異常。下面分別舉例說明。

```
>>> try:
...     10 / 0
... except:
...     print ("There' s an error.")
...
There' s an error.
>>>

>>> try:
...     10/0
... except Exception as e:
...     print (e)
...
integer division or modulo by zero
>>>
```

當單獨使用 except、不捕捉任何異常時，可以確保程式在遭遇任何異常時都不會被中斷，並傳回自訂列印出來的錯誤訊息（例如上面實例中的 There' s an error.）來代替所有可能出現的異常類型。這麼做的優點是省事，但缺點也很明顯：我們無法知道程式中實際出現了什麼類型的異常，導致程式校正困難。

而使用 Exceptions 時（except Exception as e:），不僅可以捕捉除 SystemExit、KeyboardInterrupt、GeneratorExit 外的所有異常，還能讓 Python 告訴我們實際的錯誤原因（例如上面實例中出現的除零錯誤），方便對程式進行校正。如果要捕捉 SystemExit、KeyboardInterrupt 和 GeneratorExit 這 3 種異常，則只需要把 Exceptions 取代成 BaseExceptions 即可。

Python 網路運行維護實驗 （GNS3 模擬器）

在第 2 章和第 3 章詳細介紹了 Python 的基本語法和進階語法後，本章和第 5 章將分別以實驗和實戰的形式說明 Python 在網路運行維護中的實際應用。本章共分為 4 個實驗，實驗難度循序漸進，所有實驗都將在 GNS3 模擬器上示範。本章和第 5 章的每個 Python 指令稿程式都將提供詳細的分段說明，並且提供指令稿執行前、指令稿執行中、指令稿執行後的畫面，幫助讀者清晰、直觀地了解 Python 是如何把繁雜、單調、耗時的傳統網路運行維護工作實現自動化的。

4.1 實驗執行環境

❏ 模擬器執行環境

主機作業系統：Windows 10 上執行 CentOS 8（VMware 虛擬機器）

網路裝置：GNS3 模擬器上執行的思科三層交換機

網路裝置 OS 版本：思科 IOS（vios_12-ADVENTERPRISEK9-M）

Python 版本：3.8.2

實驗網路拓撲圖如下所示。

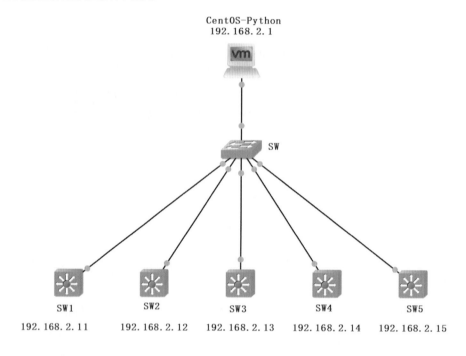

區域網 IP 位址段：192.168.2.0 /24

執行 Python 的 CentOS 主機：192.168.2.1

SW1：192.168.2.11

SW2：192.168.2.12

SW3：192.168.2.13

SW4：192.168.2.14

SW5：192.168.2.15

所有交換機都已經預先設定好了 SSH 和 Telnet，使用者名稱為 python，密碼為 123，使用者的許可權為 15 級，SSH 或 Telnet 登入後直接進入特權模式，無須輸入 enable 密碼。

4.2 Python 中的 Telnet 和 SSH 模組

在 Python 中，支援 Telnet/SSH 遠端登入存取網路裝置的模組很多，常見的有 Telnetlib、Ciscolib、Paramiko、Netmiko 和 Pexpect。其中，Telnetlib 和 Ciscolib 對應 Telnet 協定，後面 3 個對應 SSH 協定。

因為篇幅有限，Ciscolib 和 Pexpect 不在本書的討論範圍之內，本書將簡單介紹 Telnetlib 和 Netmiko，重點介紹 Paramiko，後面的實驗指令稿也都將以 Paramiko 為基礎完成。關於這 3 種模組的使用方法將透過下面的實驗來介紹。

實驗目的：透過 Telnetlib、Netmiko 和 Paramiko 模組，分別登入交換機 SW1（192.168.2.11）、SW2（192.168.2.12）、SW3（192.168.2.13），給 SW1 的 loopback 1 通訊埠設定 IP 位址 1.1.1.1/32，給 SW2 的 loopback 1 通訊埠設定 IP 位址 2.2.2.2/32，給 SW3 的 loopback 1 通訊埠設定 IP 位址 3.3.3.3/32。

4.2.1 Telnetlib

在 Python 中，我們使用 Telnetlib 模組來 Telnet 遠端登入網路裝置，Telnetlib 為 Python 內建模組，不需要 pip 下載安裝就能直接使用。鑑於 Telnet 的安全性，通常不建議在生產網路中使用 Telnet，這裡只舉一例來說明 Telnetlib 模組的使用方法。

Telnetlib 在 Python 2 和 Python 3 中有非常大的區別，雖然本書內容基於 Python 3.8，但是鑑於部分讀者對 Telnet 還有需求，並且可能使用過 Python 2，這裡將分別介紹 Telnetlib 在 Python 2 和 Python 3 中的使用方法。

1. Telnetlib 在 Python 2 中的應用

首先手動登入 SW1，確認它此時沒有 loopback 1 這個通訊埠，再開啟 debug telnet 便於後面執行程式時進行驗證。

```
SW1#
SW1#show run int loop 1
                  ^
% Invalid input detected at '^' marker.

SW1#debug telnet
Incoming Telnet debugging is on
SW1#
```

然後在執行 Python 的主機上（後面統稱「主機」）建立下面的指令稿，將其命名為 telnet.py。

```
[root@CentOS-Python ~]# cat telnet.py
import telnetlib

host = "192.168.2.11"
user = "python"
password = "123"

tn = telnetlib.Telnet(host)
tn.read_until("Username: ")
tn.write(user + "\n")
tn.read_until("Password: ")
tn.write(password + "\n")

tn.write("conf t\n")
tn.write("int loopback 1\n")
tn.write("ip address 1.1.1.1 255.255.255.255\n")
tn.write("end\n")
tn.write("exit\n")

print (tn.read_all())
```

程式分段說明如下。

（1）首先透過 import 敘述匯入 telnetlib 模組。

```
import telnetlib
```

（2）然後建立 host、user、password 3 個變數，分別對應 SW1 的管理 IP 位址、Telnet 使用者名稱和密碼，注意這 3 個變數的資料類型均為字串。

```
host = "192.168.2.11"
user = "python"
password = "123"
```

（3）呼叫 telnetlib 的 Telnet() 函數，將它設定值給變數 tn，嘗試以 Telnet 方式登入 192.168.2.11。

```
tn = telnetlib.Telnet(host)
```

（4）透過 Telnet 登入思科交換機時，終端輸出最下面提示的資訊始終為 "Username:"，如下圖所示。

```
[root@localhost ~]# telnet 192.168.2.11
Trying 192.168.2.11...
Connected to 192.168.2.11.
Escape character is '^]'.

**********************************************************************
* IOSv - Cisco Systems Confidential                                  *
*                                                                    *
* This software is provided as is without warranty for internal      *
* development and testing purposes only under the terms of the Cisco  *
* Early Field Trial agreement.  Under no circumstances may this software *
* be used for production purposes or deployed in a production         *
* environment.                                                        *
*                                                                    *
* By using the software, you agree to abide by the terms and conditions *
* of the Cisco Early Field Trial Agreement as well as the terms and    *
* conditions of the Cisco End User License Agreement at               *
* http://www.cisco.com/go/eula                                        *
*                                                                    *
* Unauthorized use or distribution of this software is expressly      *
* Prohibited.                                                         *
**********************************************************************
User Access Verification

Username:
```

我們透過 tn.read_until("Username:") 函數來告訴 Python：如果在終端資訊
裡讀到 "Username:" 字樣，則使用 tn.write(user + "\n") 函數來輸入 Telnet 使
用者名稱並確認。

```
tn.read_until("Username: ")
tn.write(user + "\n")
```

（5）同理，在輸入 Username 後，接下來終端將顯示 "Password:"，如下圖
所示。

```
[root@localhost ~]# telnet 192.168.2.11
Trying 192.168.2.11...
Connected to 192.168.2.11.
Escape character is '^]'.

**********************************************************************
* IOSv - Cisco Systems Confidential                                  *
*                                                                    *
* This software is provided as is without warranty for internal     *
* development and testing purposes only under the terms of the Cisco *
* Early Field Trial agreement.  Under no circumstances may this software *
* be used for production purposes or deployed in a production       *
* environment.                                                       *
*                                                                    *
* By using the software, you agree to abide by the terms and conditions *
* of the Cisco Early Field Trial Agreement as well as the terms and *
* conditions of the Cisco End User License Agreement at             *
* http://www.cisco.com/go/eula                                       *
*                                                                    *
* Unauthorized use or distribution of this software is expressly    *
* Prohibited.                                                        *
**********************************************************************

User Access Verification

Username: python
Password:
```

我們透過 tn.read_until("Password: ") 函數來告訴 Python：如果在終端資
訊裡讀到 "Password:" 字樣，則使用 tn.write(password + "\n") 函數來輸入
Telnet 密碼並確認。

```
tn.read_until("Password: ")
tn.write(password + "\n")
```

（6）至此，我們已經讓 Python 成功透過 Telnet 登入 SW1 了。接下來就是繼續用 Telnetlib 的 write() 函數在 SW1 上輸入各種設定指令，給 loopback 1 通訊埠設定 1.1.1.1/32 這個 IP 位址，這是網路工程師最熟悉的部分，就不再解釋了。這裡只提一點，必須透過 tn.write ("exit\n") 來退出 Telnet，否則位於指令稿尾端的 print tn.read_all()將故障，原因在下一步中解釋。

```
tn.write("conf t\n")
tn.write("int loopback 1\n")
tn.write("ip address 1.1.1.1 255.255.255.255\n")
tn.write("end\n")
tn.write("exit\n")
```

（7）最後我們透過 Telnetlib 的 read_all() 方法將登入 SW1 後執行指令的所有過程都記錄下來，透過 print (tn.read_all()) 將其列印，這樣就能清楚地看到 Python 對 SW1 做了什麼。註：read_all() 方法只有在退出 Telnet 後才會生效，因此我們必須在其之前透過 tn.write ("exit\n") 退出 Telnet。

```
print (tn.read_all())
```

執行程式後看結果、做驗證。

（1）在主機上輸入 python2 telnet.py，用 Python 2 來執行程式，如下圖所示。

```
[root@CentOS-Python ~]# python2 telnet.py

***************************************************************
* IOSv is strictly limited to use for evaluation, demonstration and IOS  *
* education. IOSv is provided as-is and is not supported by Cisco's  *
* Technical Advisory Center. Any use or disclosure, in whole or in part, *
* of the IOSv Software or Documentation to any third party for any  *
* purposes is expressly prohibited except as otherwise authorized by  *
* Cisco in writing.  *
***************************************************************
SW1#conf t
Enter configuration commands, one per line.  End with CNTL/Z.
SW1(config)#int loopback 1
SW1(config-if)#ip address 1.1.1.1 255.255.255.255
SW1(config-if)#end
SW1#exit

[root@CentOS-Python ~]#
```

（2）同一時間在 SW1 上可以看到 Python 透過 Telnet 登入 SW 的詳細
debug 記錄檔（我們已經在 SW1 上開啟了 debug telnet），這裡能夠看
到 "*May　2 14:00: 10.023:%LINK-3- UPDOWN: Interface Loopback1,
changed state to up"、"*May　2 14:00: 10.668: %SYS-5- CONFIG_I:
Configured from console by python on vty0 (192.168.2.1)" 和 "*May 2 14:00:
11.060:%LINEPROTO-5-UPDOWN: Line protocol on Interface Loopback1,
changed state to up"3 筆記錄檔，表明 loopback 1 通訊埠已經被執行 Python
的主機 192.168.2.1 建立，並且通訊埠已經被開啟了。

```
SW1#
*May  2 14:00:07.396: Telnet2: 1 1 251 1
*May  2 14:00:07.397: TCP2: Telnet sent WILL ECHO (1)
*May  2 14:00:07.397: Telnet2: 2 2 251 3
*May  2 14:00:07.398: TCP2: Telnet sent WILL SUPPRESS-GA (3)
*May  2 14:00:07.398: Telnet2: 80000 80000 253 24
*May  2 14:00:07.399: TCP2: Telnet sent DO TTY-TYPE (24)
*May  2 14:00:07.399: Telnet2: 10000000 10000000 253 31
*May  2 14:00:07.400: TCP2: Telnet sent DO WINDOW-SIZE (31)
*May  2 14:00:07.466: TCP2: Telnet received DONT ECHO (1)
*May  2 14:00:07.467: TCP2: Telnet sent WONT ECHO (1)
*May  2 14:00:07.479: TCP2: Telnet received DONT SUPPRESS-GA (3)
*May  2 14:00:07.479: TCP2: Telnet sent WONT SUPPRESS-GA (3)
SW1#
*May  2 14:00:07.481: TCP2: Telnet received WONT TTY-TYPE (24)
*May  2 14:00:07.482: TCP2: Telnet sent DONT TTY-TYPE (24)
*May  2 14:00:07.484: TCP2: Telnet received WONT WINDOW-SIZE (31)
*May  2 14:00:07.486: TCP2: Telnet sent DONT WINDOW-SIZE (31)
*May  2 14:00:07.663: TCP2: Telnet received DONT ECHO (1)
*May  2 14:00:07.664: TCP2: Telnet received DONT SUPPRESS-GA (3)
*May  2 14:00:07.664: TCP2: Telnet received WONT TTY-TYPE (24)
*May  2 14:00:07.664: TCP2: Telnet received WONT WINDOW-SIZE (31)
SW1#
*May  2 14:00:10.023: %LINK-3-UPDOWN: Interface Loopback1, changed state to up
SW1#
*May  2 14:00:10.668: %SYS-5-CONFIG_I: Configured from console by python on vty0 (192.168.2.1)
*May  2 14:00:11.060: %LINEPROTO-5-UPDOWN: Line protocol on Interface Loopback1, changed state to up
SW1#
```

（3）最後在 SW1 上輸入 show run int loop 1 做驗證，確認 loopback 1 通訊
埠的 IP 位址設定正確，透過 Telnetlib 模組登入交換機修改設定的實驗成
功，如下圖所示。

```
SW1#show run int loop 1
Building configuration...

Current configuration : 63 bytes
!
interface Loopback1
 ip address 1.1.1.1 255.255.255.255
end

SW1#
```

2. Telnetlib 在 Python 3 中的應用

在 Python 2 中，Telnetlib 模組下所有函數的傳回值均為字串，例如前面列出的程式中的 tn.read_until("Username: ")，這裡的 "Username:" 即一個字串。同理，tn.write(user + "\n") 中的 user 雖然是一個變數，但是該變數的類型依然為字串，後面的 tn.read_until ("Password: ") 和 tn.write(password + "\n")，以及 tn.write('conf t\n') 等都是同樣的原理。

而在 Python 3 中，Telnetlib 模組下所有函數的傳回值都變成位元組字串（Byte Strings）。因此，在 Python 3 中使用 Telnetlib 需要注意以下幾點。

- 在字串的前面需要加上一個 b。
- 在變數和 Telnetlib 函數後面需要加上 .encode('ascii') 函數。
- 在 read_all() 函數後面需要加上 decode('ascii') 函數。

下面是在 Python 3 或 Python 3.8 中使用的 Telnetlib 指令稿。

```
import telnetlib

host = "192.168.2.11"
user = "python"
password = "123"

tn = telnetlib.Telnet(host)
tn.read_until(b"Username: ")
tn.write(user.encode('ascii') + b"\n")
```

```
tn.read_until(b"Password: ")
tn.write(password.encode('ascii') + b"\n")

tn.write(b"conf t\n")
tn.write(b"int loopback 1\n")
tn.write(b"ip address 1.1.1.1 255.255.255.255\n")
tn.write(b"end\n")
tn.write(b"exit\n")

print (tn.read_all().decode('ascii'))
```

值得一提的是，這段 Python 3 的 Telnetlib 指令稿在 Python 2 中依然可以
正確使用。但是 Python 2 的那段指令稿在 Python 3 中卻無法相容，有興
趣的讀者可以自行在 Python 3 或 Python 3.8 中試驗一下，如果不使用 b
和 .encode('ascii') 函數來做位元組字串會收到什麼類型的錯誤。

4.2.2 Paramiko 和 Netmiko

Python 中支援 SSH 協定實現遠端連接裝置的模組主要有 Paramiko 和
Netmiko 兩種。Paramiko 是 Python 中一個非常著名的開放原始碼 SSHv2
專案，基於 Python 2.7 和 Python 3.4+ 開發，於 2013 年左右發佈，最早的
作者是 Jeff Forcier。Paramiko 同時支援 SSH 的服務端和用戶端，原始程式
可以在 GitHub 上下載。

Netmiko 是另一個 SSH 開放原始碼專案，於 2014 年年底發佈，作者是
Kirk Byers。根據作者的介紹，Netmiko 是以 Paramiko 專案開發為基礎
的，在 Paramiko 的基礎上主要做了支援多廠商裝置、簡化指令 show 的執
行和回應內容的讀取、簡化網路裝置的設定指令等改進，實際的改進細節
在後面會提到。

目前，NetDevOps 界的主流意見是 Netmiko 比 Paramiko 好用，事實上也
的確如此。但是根據筆者學習和使用 Python 的經驗來看，Netmiko 將太多

東西簡化、最佳化，反而不利於初學者學習。初學者如果從 Paramiko 上手，則能更直觀地了解使用 Python 來 SSH 遠端登入、管理裝置時需要注意什麼問題，更全面地鍛鍊自己的程式設計能力，並且在需要時能更順利地在短時間內上手 Netmiko，讓自己的技術更全面。反之則基本是不可逆的（可以想像成從一開始就學手排汽車和從一開始就學自排汽車的兩位駕駛的區別）。

基於上述考慮，本章和第 5 章的所有實驗都將以 Paramiko 模組實現，在第 6 章中會說明一些以 Netmiko 為基礎的 Python 協力廠商模組在網路運行維護中的應用，這裡只以一個簡單的實驗來說明如何使用 Netmiko 登入 SW2，並為它的 loopback 1 通訊埠設定 IP 位址 2.2.2.2/32。

1. Netmiko 實驗舉例

Netmiko 為 Python 協力廠商模組，需要使用 pip 來下載安裝，因為實驗基於 Python 3.8，這裡用指令 pip3.8 install Netmiko 來下載安裝 Netmiko，如下圖所示。

下載完畢後，進入 Python 3.8 解譯器，如果 import netmiko 沒有顯示出錯，則説明 Netmiko 安裝成功，如下圖所示。

```
[root@CentOS-Python ~]# python3.8
Python 3.8.2 (default, Apr 27 2020, 23:06:10)
[GCC 8.3.1 20190507 (Red Hat 8.3.1-4)] on linux
Type "help", "copyright", "credits" or "license" for more information.
>>> import netmiko
>>>
```

與 Telnetlib 的實驗步驟一樣，首先登入 SW2，確認 loopback 1 通訊埠目前不存在，然後啟用 debug ip ssh 來監督驗證下面建立的 Netmiko 指令稿是否以 SSH 協定登入存取 SW2，如下圖所示。

```
SW2#show run int loop1
                   ^
% Invalid input detected at '^' marker.

SW2#debug ip ssh
Incoming SSH debugging is on
SW2#
```

接下來在主機上建立一個名為 ssh_Netmiko.py 的 Python 3.8 指令稿，內容如下。

```
[root@CentOS-Python ~]# cat ssh_Netmiko.py
from netmiko import ConnectHandler

SW2 = {
    'device_type': 'cisco_ios',
    'ip': '192.168.2.12',
    'username': 'python',
    'password': '123',
}

connect = ConnectHandler(**SW2)
print ("Successfully connected to " + SW2['ip'])
config_commands = ['int loop 1', 'ip address 2.2.2.2 255.255.255.255']
output = connect.send_config_set(config_commands)
print (output)
```

```
result = connect.send_command('show run int loop 1')
print (result)
```

程式分段說明如下。

（1）首先透過 import 敘述從 Netmiko 模組匯入它的程式庫函數 ConnectHandler()。該函數用來實現 SSH 登入網路裝置，是 Netmiko 最重要的函數。

```
from Netmiko import ConnectHandler
```

（2）建立一個名為 SW2 的字典，該字典包含 "device_type"、"ip"、"username" 和 "password" 4 個必選的鍵，其中後面 3 個鍵的意思很好了解，這裡主要說明 "device_type"。前面提到支援多廠商的裝置是 Netmiko 的優勢之一，截至 2019 年 1 月，Netmiko 支援 Arista、Cisco、HP、Juniper、Alcatel、Huawei 和 Extreme 等絕大多數主流廠商的裝置。除此之外，Netmiko 同樣支援擁有多種不同 OS 類型的廠商的裝置，例如針對 Cisco 的裝置，Netmiko 能同時支援 Cisco ASA、Cisco IOS、Cisco IOS-XE、Cisco IOS-XR、Cisco NX-OS 和 Cisco SG300 共 6 種不同 OS 類型的裝置。由於不同廠商的裝置登入 SSH 後命令列介面和特性不盡相同，因此我們必須透過 "device_type" 來指定需要登入的裝置的類型。因為實驗裡我們用到的是 Cisco IOS 裝置，因此 "deivce_type" 的鍵值為 "cisco_ios"。

```
SW2 = {
    'device_type': 'cisco_ios',
    'ip': '192.168.2.12',
    'username': 'python',
    'password': '123',
}
```

（3）呼叫 ConnectHandler() 函數，用已經建立的字典 SW2 進行 SSH 連接，將它設定值給 connect 變數，注意 SW2 前面的 ** 作為標識，不可以省去。

```
connect = ConnectHandler(**SW2)
```

（4）如果 SSH 登入裝置成功，則提示使用者並告知所登入的交換機的 IP
位址。

```
print ("Successfully connected to " + SW2['ip'])
```

（5）建立一個名為 config_commands 的串列，其元素為需要在交換機上依
次執行的指令。

```
config_commands = ['int loop 1' , 'ip address 2.2.2.2 255.255.255.255']
```

（6）然後以剛剛建立的 config_commands 串列為參數，呼叫
ConnectHandler() 的 send_config_set() 函數來使用上述指令對 SW2 做設
定，並將設定過程列印出來。

```
output = connect.send_config_set(config_commands)
print (output)
```

（7）最後呼叫 ConnectHandler() 的 send_command() 函數，對交換機輸入
指令 show run int loop 1 並將回應內容列印出來。需要注意的是，send_
command() 一次只能向裝置輸入一個指令，而 send_config_set() 則可向裝
置一次輸入多個指令。

```
result = connect.send_command('show run int loop 1')
print (result)
```

執行程式後看結果、做驗證。

（1）在主機上輸入 python3.8 ssh_Netmiko.py 執行程式，可以看到除了我
們在程式裡寫的 int loop 1 和 ip address 2.2.2.2 255.255.255.255 兩個指令，
Netmiko 額外替我們輸入了 3 個指令，一個是 config term，一個是 end，
還有一個是 write memory。最後我們能看到 show run int loop 1 指令的回
應內容，證實 loopback1 通訊埠設定成功，如下圖所示。

```
[root@CentOS-Python ~]# python3.8 ssh_netmiko.py
Sucessfully connected to 192.168.2.12
config term
Enter configuration commands, one per line.  End with CNTL/Z.
SW2(config)#int loop 1
SW2(config-if)#ip address 2.2.2.2 255.255.255.255
SW2(config-if)#end
SW2#

Building configuration...

Current configuration : 63 bytes
!
interface Loopback1
 ip address 2.2.2.2 255.255.255.255
end

[root@CentOS-Python ~]#
```

（2）同一時間在 SW2 上可以看到透過 Netmiko 來 SSH 登入 SW2 的詳細記錄檔（我們已經在 SW2 上開啟了 debug ip ssh），如下圖所示。

```
SW2#
*May  3 05:25:13.262: SSH0: starting SSH control process
*May  3 05:25:13.262: SSH0: sent protocol version id SSH-2.0-Cisco-1.25
*May  3 05:25:13.275: SSH0: protocol version id is - SSH-2.0-paramiko_2.7.1
*May  3 05:25:13.275: SSH2 0: kexinit sent: hostkey algo = ssh-rsa
*May  3 05:25:13.276: SSH2 0: kexinit sent: encryption algo = aes128-ctr,aes192-ctr,aes256-ctr,aes128-cbc,3des-cbc,aes192
-cbc,aes256-cbc
*May  3 05:25:13.292: SSH2 0: kexinit sent: mac algo = hmac-sha1,hmac-sha1-96
*May  3 05:25:13.293: SSH2 0: send:packet of  length 368 (length also includes padlen of 5)
*May  3 05:25:13.297: SSH2 0: SSH2_MSG_KEXINIT sent
*May  3 05:25:13.298: SSH2 0: ssh_receive: 880 bytes received
*May  3 05:25:13.298: SSH2 0: input: total packet length of 880 bytes
*May  3 05:25:13.299: SSH2 0: partial packet length(block size)8 bytes,needed 872 bytes,
          maclen 0
```

（3）最後我們在 SW2 上輸入 show run int loop 1 做驗證，確認 loopback 1 通訊埠的 IP 位址設定正確，透過 Netmiko 模組登入交換機修改設定的實驗成功，如下圖所示。

```
SW2#show run int loop1
Building configuration...

Current configuration : 63 bytes
!
interface Loopback1
 ip address 2.2.2.2 255.255.255.255
end

SW2#
```

註：在使用 Netmiko 的 ConnectHandler 的指令稿中，不能將指令稿命名為 netmiko.py，否則會遇到 "ImportError: cannot import name 'ConnectHandler' from partially initialized module 'Netmiko' (most likely due to a circular import) (/root/Netmiko.py)" 這個錯誤。Paramiko 也一樣，指令稿名字不能為 paramiko.py，如下圖所示。

```
[root@CentOS-Python ~]# python3.8 netmiko.py
Traceback (most recent call last):
  File "netmiko.py", line 1, in <module>
    from netmiko import ConnectHandler
  File "/root/netmiko.py", line 1, in <module>
    from netmiko import ConnectHandler
ImportError: cannot import name 'ConnectHandler' from partially initialized modu
le 'netmiko' (most likely due to a circular import) (/root/netmiko.py)
[root@CentOS-Python ~]#
```

2. Paramiko 實驗舉例

與 Netmiko 不同，Paramiko 不會在做設定的時候替我們自動加上 config term、end 和 write memory 等指令，也不會在執行各種 show 指令後自動儲存該指令的回應內容，一切都需要我們手動搞定。另外，Python 不像人類，後者在手動輸入每個指令後會間隔一定時間，再輸入下一個指令。Python 是一次性執行所有指令稿裡的指令的，中間沒有間隔時間。當你要一次性輸入很多個指令時，便經常會發生 SSH 終端跟不上速度，導致某些指令缺失沒有被輸入的問題（用傳統的「複製、貼上」方法給網路裝置做設定的人應該遇到過這個問題）。同樣，在用 print() 函數輸入回應內容或用 open.write() 將回應內容寫入文件進行儲存時，也會因為缺乏間隔時間而導致 Python「截圖」不完整，進一步導致回應內容不完整。Netmiko 自動幫我們解決了這個問題，也就是説，不管上面所舉的 Netmiko 實例中 config_commands 串列中的元素（指令）有多少個，都不會出現因為間隔時間不足而導致設定指令缺失的問題。而在 Paramiko 中，我們必須匯入 time 模組，使用該模組的 sleep() 方法來解決這個問題，關於 time 模組和 sleep() 方法的使用將在下面的實驗中講到。

下面我們用 Paramiko 來完成最後一個實驗：用 Paramiko 來登入 SW3 並為它的 loopback 1 通訊埠設定 IP 位址 3.3.3.3/32。

與 Netmiko 一樣，Paramiko 也是協力廠商模組，需要使用 pip 來下載安裝，方法如下圖所示。

```
[root@CentOS-Python ~]# pip3.8 install paramiko
Requirement already satisfied: paramiko in /usr/local/lib/python3.8/site-package
s (2.7.1)
Requirement already satisfied: bcrypt>=3.1.3 in /usr/local/lib/python3.8/site-pa
ckages (from paramiko) (3.1.7)
Requirement already satisfied: pynacl>=1.0.1 in /usr/local/lib/python3.8/site-pa
ckages (from paramiko) (1.3.0)
Requirement already satisfied: cryptography>=2.5 in /usr/local/lib/python3.8/sit
e-packages (from paramiko) (2.9.2)
Requirement already satisfied: cffi>=1.1 in /usr/local/lib/python3.8/site-packag
es (from bcrypt>=3.1.3->paramiko) (1.14.0)
Requirement already satisfied: six>=1.4.1 in /usr/local/lib/python3.8/site-packa
ges (from bcrypt>=3.1.3->paramiko) (1.14.0)
Requirement already satisfied: pycparser in /usr/local/lib/python3.8/site-packag
es (from cffi>=1.1->bcrypt>=3.1.3->paramiko) (2.20)
WARNING: You are using pip version 19.2.3, however version 20.1 is available.
You should consider upgrading via the 'pip install --upgrade pip' command.
[root@CentOS-Python ~]#
```

下載完畢後，進入 Python 3.8 解譯器，如果 import paramiko 沒有顯示出錯，則說明 Paramiko 安裝成功，如下圖所示。

```
[root@CentOS-Python ~]# python3.8
Python 3.8.2 (default, Apr 27 2020, 23:06:10)
[GCC 8.3.1 20190507 (Red Hat 8.3.1-4)] on linux
Type "help", "copyright", "credits" or "license" for more information.
>>> import paramiko
>>>
```

與前面兩個實驗一樣，首先登入 SW3，確認 loopback 1 通訊埠目前並不存在，然後啟用 debug ip ssh 來監督驗證 Paramiko 指令稿是否以 SSH 協定登入存取 SW3，如下圖所示。

```
SW3#show run int loop1
                    ^
% Invalid input detected at '^' marker.

SW3#debug ip ssh
Incoming SSH debugging is on
SW3#
```

接下來在主機上建立一個名為 ssh_paramiko.py 的 Python 3.8 指令稿，內容如下。

```python
import paramiko
import time


ip = "192.168.2.13"
username = "python"
password = "123"


ssh_client = paramiko.SSHClient()
ssh_client.set_missing_host_key_policy(paramiko.AutoAddPolicy())
ssh_client.connect(hostname=ip，username=username，password=password)

print ("Successfully connected to "，ip)
command = ssh_client.invoke_shell()
command.send("configure terminal\n")
command.send("int loop 1\n")
command.send("ip address 3.3.3.3 255.255.255.255\n")
command.send("end\n")
command.send("wr mem\n")

time.sleep(2)
output = command.recv(65535)
print (output.decode("ascii"))


ssh_client.close
```

程式分段説明如下。

（1）首先透過 import 敘述匯入 paramiko 和 time 兩個模組。

```
import paramiko
import time
```

（2）建立 3 個變數：ip、username 和 password，分別對應我們要登入的交換機（SW3）的管理 IP 位址、SSH 使用者名稱和密碼。

```
ip = "192.168.2.13"
username = "python"
password = "123"
```

（3）呼叫 Paramiko 的 SSHClient() 方法，將其設定值給變數 ssh_client。顧名思義，CentOS 主機做 SSH 用戶端，而 SSH 服務端則是我們要登入的 SW3（192.168.2.13）。

```
ssh_client = paramiko.SSHClient()
```

（4）預設情況下，Paramiko 會拒絕任何未知的 SSH 公開金鑰（publickey），這裡我們需要使用 ssh_client.set_missing_host_key_policy(paramiko.AutoAddPolicy()) 來讓 Paramiko 接受 SSH 服務端（也就是 SW3）提供的公開金鑰，這是任何時候使用 Paramiko 都要用到的標準設定。

```
ssh_client.set_missing_host_key_policy(paramiko.AutoAddPolicy())
```

（5）在做完 Paramiko 關於 SSH 公開金鑰相關的設定後，呼叫 Paramiko.SSHClient() 的 connect()函數進行 SSH 登入，該函數包含 3 個必選的參數 hostname、username 和 password，分別對應我們建立的 ip、username 和 password 3 個變數，也就是遠端登入的裝置的主機名稱 /IP 位址、SSH 使用者名稱和密碼。如果 SSH 登入裝置成功，則提示使用者並告知所登入的交換機的管理 IP 位址。

```
ssh_client.connect (hostname=ip，username=username，password=password)
print  ("Successfully connected to "，ip)
```

（6）SSH 連接成功後，需要呼叫 Paramiko.SSHClient() 的 invoke_shell()
方法來喚醒 shell，也就是思科裝置 IOS 命令列，並將它設定值給變數
command。

```
command = ssh_client.invoke_shell()
```

（7）之後便可以呼叫 invoke_shell() 的 command() 函數來向 SW3「發號施
令」了。這裡注意需要手動在程式中輸入 configure terminal、end 和 wr
mem 3 個指令，這一點與 Netmiko 不同。

```
command.send("configure terminal\n")
command.send("int loop 1\n")
command.send("ip address 3.3.3.3 255.255.255.255\n")
command.send("end\n")
command.send("wr mem\n")
```

（8）前面有講到，Python 是一次性執行所有指令稿裡的指令，中間沒
有間隔時間，這樣會導致某些指令遺漏和回應內容不完整的問題。在用
Paramiko 的 recv() 函數將回應結果儲存之前，我們需要呼叫 time 模組下
的 sleep() 函數手動讓 Python 休眠 2s，這樣回應內容才能被完整列印出來
（sleep() 中參數的單位為 s）。這裡的 command.recv(65535) 中的 65535 代
表截取 65535 個字元的回應內容，這也是 Paramiko 一次能截取的最大回
應內容數。另外，與 Telnetlib 類似，在 Python 3 中，Paramiko 截取的回應
內容格式為位元組字串，需要用 decode("ascii") 將其解析為 ASCII 編碼，
否則列印出來的 output 的內容格式會很難看（下面驗證部分會列出不帶
decode("ascii") 的回應內容供參考）。

```
time.sleep(2)
output = command.recv(65535)
print (output.decode("ascii"))
```

（9）設定完畢後，使用 close 方法退出 SSH。

```
ssh_client.close
```

執行程式後看結果、做驗證。

（1）在主機上輸入 python ssh_paramiko.py 來執行程式，可以看到指令稿提示登入 192.168.2.13（SW3）成功，然後給 loopback1 通訊埠設定 IP 位址 3.3.3.3，最後儲存設定。程式執行如下圖所示。

上面提到，如果我們不在 print(output) 後面加上 decode("ascii") 會怎樣呢？如下圖所示。

是不是很亂？根本不知道指令稿在做什麼。

註：如果在 CentOS 中執行程式時遇到 SSH 使用者名稱和密碼都正確，但是一直出現 "Authentication failed" 驗證失敗的異常提示（如下圖所示），很可能是主機之前用 ssh-keygen 指令產生過 RSA 金鑰對用來做 SSH 免密碼登入（例如在使用 Gitlab 時可以用到）。一旦在 CentOS 上產生了本機 RSA 金鑰對，Paramiko 就會一直嘗試使用該金鑰對來登入裝置，原因是預設情況下 Paramiko.SSHClient().connect() 中的 look_for_keys 可選參數預設為 True。大部分的情況下，我們登入交換機等網路裝置時不是 SSH 免密登入，沒有用到 RSA 金鑰對，但是 Paramiko 又一直嘗試使用該金鑰對來登入裝置，導致 "Authentication failed" 驗證失敗。

```
[root@localhost ~]# python ssh_paramiko.py
Traceback (most recent call last):
  File "ssh_paramiko.py", line 10, in <module>
    ssh_client.connect(hostname=ip,username=username,password=password,look_for_keys=True)
  File "/usr/lib/python2.7/site-packages/paramiko/client.py", line 437, in connect
    passphrase,
  File "/usr/lib/python2.7/site-packages/paramiko/client.py", line 749, in _auth
    raise saved_exception
paramiko.ssh_exception.AuthenticationException: Authentication failed.
[root@localhost ~]#
```

在交換機上開啟 debug ip ssh 後，能看到使用者名稱 python 的公開金鑰缺失，導致驗證失敗的記錄檔記錄，如下圖所示。

```
*Mar 25 06:52:18.152: SSH2 0: Using method = publickey
*Mar 25 06:52:18.153: S
S3#SH2 0: Publickey for 'python' not found
*Mar 25 06:52:18.153: SSH2 0: Pubkey Authentication failed for user 'python'
*Mar 25 06:52:18.153: SSH0: password authentication failed for python
*Mar 25 06:52:20.153: SSH2 0: Authentications that can continue = publickey,keyboard-interactive,password
*Mar 25 06:52:20.153: SSH2 0: send:packet of  length 64 (length also includes padlen of 14)
*Mar 25 06:52:20.153: SSH2 0: computed MAC for sequence no.#6 type 51
*Mar 25 06:52:20.169: SSH2 0: ssh_receive: 52 bytes received
*Mar 25 06:52:20.169: SSH2 0: input: total packet length of 32 bytes
*Mar 25 06:52:20.169: SSH2 0: partial packet length(block size)16 bytes,needed 16 bytes,
                maclen 20
*Mar 25 06:52:20.170: SSH2 0: MAC compared for #6 :ok
*Mar 25 06:52:20.170: SSH2 0: input: padlength 10 bytes
S3#
*Mar 25 06:52:20.171: SSH2 0: send:packet of  length 80 (length also includes padlen of 16)
*Mar 25 06:52:20.171: SSH2 0: computed MAC for sequence no.#7 type 1
*Mar 25 06:52:20.274: SSH0: Session disconnected - error 0x07
S3#
```

解決辦法也很簡單，將 look_for_keys 參數修改為 False 即可。

```
ssh_client.connect(hostname=ip，username=username，password=password，
look_for_keys=False)
```

（2）同一時間在 SW3 上可以看到我們透過 Python 來 SSH 登入 SW3 的詳
細記錄檔（之前我們已經在 SW3 開啟了 debug ip ssh），如下圖所示。

```
SW3#debug ip ssh
Incoming SSH debugging is on
SW3#
*May  3 05:48:26.936: SSH0: starting SSH control process
*May  3 05:48:26.937: SSH0: sent protocol version id Cisco-1.25
*May  3 05:48:26.973: SSH0: protocol version id is - SSH-2.0-paramiko_2.7.1
*May  3 05:48:26.975: SSH2 0: kexinit sent: hostkey algo = ssh-rsa
*May  3 05:48:26.976: SSH2 0: kexinit sent: encryption algo = aes128-ctr,aes192-ctr,aes256-ctr,aes128-cbc,3des-cbc,aes192
-cbc,aes256-cbc
*May  3 05:48:26.977: SSH2 0: kexinit sent: mac algo = hmac-sha1,hmac-sha1-96
*May  3 05:48:26.979: SSH2 0: send:packet of  length 368 (length also includes padlen of 5)
*May  3 05:48:26.989: SSH2 0: SSH2_MSG_KEXINIT sent
*May  3 05:48:26.991: SSH2 0: ssh_receive: 880 bytes received
*May  3 05:48:26.992: SSH2 0: input: total packet length of 880 bytes
*May  3 05:48:26.993: SSH2 0: partial packet length(block size)8 bytes,needed 872 bytes,
           maclen 0
*May  3 05:48:26.993: SSH2 0: input: padlength 10 bytes
```

（3）最後我們在 SW3 上輸入 show run int loop 1 做驗證，確認 loopback 1
通訊埠的 IP 位址設定正確，透過 Paramiko 模組登入交換機修改設定的實
驗成功，如下圖所示。

```
SW3#show run int loop1
Building configuration...

Current configuration : 63 bytes
!
interface Loopback1
 ip address 3.3.3.3 255.255.255.255
end

SW3#
```

在舉例介紹了 Telnetlib、Netmiko 和 Paramiko 的基本使用方法後，下面將
正式進入實驗部分。每個實驗都將以網路運行維護中常見的需求為實驗
背景，說明怎樣用 Python 來實現這些需求的自動化（所有實驗都將使用
Paramiko 實現）。

4.3 實驗 1：input() 函數和 getpass模組

在針對 Telnetlib、Netmiko 和 Paramiko 模組的基礎知識説明中，我們都將 SSH 登入的使用者名稱和密碼明文寫在了指令稿裡，這種做法在實驗練習中可以使用，但是在生產環境中是不夠安全的。在生產環境中，正確的做法是使用 input() 函數和 getpass 模組來分別提示使用者手動輸入 SSH 使用者名稱和密碼，這是本實驗將重點説明的部分。

4.3.1 實驗目的

- 使用 input() 函數和 getpass 模組實現互動式的 SSH 使用者名稱和密碼輸入。
- 透過 for 循環同時為 5 台交換機 SW1 ～ SW5 設定 VLAN 10 ～ VLAN 20。

4.3.2 實驗準備

（1）執行程式前，首先檢查 5台交換機的設定，確認它們都沒有 VLAN 10 ～ VLAN 20。

執行程式前，SW1 的設定如下圖所示。

```
SW1#show vlan b

VLAN Name                             Status    Ports
---- -------------------------------- --------- -------------------------------
1    default                          active    Gi0/0, Gi0/1, Gi0/2, Gi0/3
                                                Gi1/0, Gi1/1, Gi1/2, Gi1/3
                                                Gi2/0, Gi2/1, Gi2/2, Gi2/3
                                                Gi3/0, Gi3/1, Gi3/2, Gi3/3
1002 fddi-default                     act/unsup
1003 token-ring-default               act/unsup
1004 fddinet-default                  act/unsup
1005 trnet-default                    act/unsup
SW1#
```

執行程式前，SW2 的設定如下圖所示。

```
SW2#show vlan b

VLAN Name                             Status    Ports
---- -------------------------------- --------- -------------------------------
1    default                          active    Gi0/0, Gi0/1, Gi0/2, Gi0/3
                                                Gi1/0, Gi1/1, Gi1/2, Gi1/3
                                                Gi2/0, Gi2/1, Gi2/2, Gi2/3
                                                Gi3/0, Gi3/1, Gi3/2, Gi3/3
1002 fddi-default                     act/unsup
1003 token-ring-default               act/unsup
1004 fddinet-default                  act/unsup
1005 trnet-default                    act/unsup
SW2#
```

執行程式前，SW3 的設定如下圖所示。

```
SW3#show vlan b

VLAN Name                             Status    Ports
---- -------------------------------- --------- -------------------------------
1    default                          active    Gi0/0, Gi0/1, Gi0/2, Gi0/3
                                                Gi1/0, Gi1/1, Gi1/2, Gi1/3
                                                Gi2/0, Gi2/1, Gi2/2, Gi2/3
                                                Gi3/0, Gi3/1, Gi3/2, Gi3/3
1002 fddi-default                     act/unsup
1003 token-ring-default               act/unsup
1004 fddinet-default                  act/unsup
1005 trnet-default                    act/unsup
SW3#
```

執行程式前，SW4 的設定如下圖所示。

```
SW4#show vlan b

VLAN Name                             Status    Ports
---- -------------------------------- --------- -------------------------------
1    default                          active    Gi0/0, Gi0/1, Gi0/2, Gi0/3
                                                Gi1/0, Gi1/1, Gi1/2, Gi1/3
                                                Gi2/0, Gi2/1, Gi2/2, Gi2/3
                                                Gi3/0, Gi3/1, Gi3/2, Gi3/3
1002 fddi-default                     act/unsup
1003 token-ring-default               act/unsup
1004 fddinet-default                  act/unsup
1005 trnet-default                    act/unsup
SW4#
```

執行程式前，SW5 的設定如下圖所示。

```
SW5#show vlan b

VLAN Name                             Status    Ports
---- -------------------------------- --------- -------------------------------
1    default                          active    Gi0/0, Gi0/1, Gi0/2, Gi0/3
                                                Gi1/0, Gi1/1, Gi1/2, Gi1/3
                                                Gi2/0, Gi2/1, Gi2/2, Gi2/3
                                                Gi3/0, Gi3/1, Gi3/2, Gi3/3
1002 fddi-default                     act/unsup
1003 token-ring-default               act/unsup
1004 fddinet-default                  act/unsup
1005 trnet-default                    act/unsup
SW5#
```

（2）在主機上建立實驗 1 的 Python 指令稿，將其命名為 lab1.py。

```
[root@CentOS-Python ~]# vi lab1.py
```

4.3.3 實驗程式

將下列程式寫入指令稿 lab1.py。

```python
import paramiko
import time
import getpass

username = input('Username: ')
password = getpass.getpass('Password: ')

for i in range(11,16):
    ip = "192.168.2." + str(i)
    ssh_client = paramiko.SSHClient()
    ssh_client.set_missing_host_key_policy(paramiko.AutoAddPolicy())
    ssh_client.connect(hostname=ip,username=username,password=password)
    print ("Successfully connect to ", ip)
    command = ssh_client.invoke_shell()
    command.send("configure terminal\n")
    for n in range (10,21):
        print ("Creating VLAN " + str(n))
```

```
        command.send("vlan " + str(n) +  "\n")
        command.send("name Python_VLAN " + str(n) +  "\n")
        time.sleep(1)

    command.send("end\n")
    command.send("wr mem\n")
    time.sleep(2)
    output = command.recv(65535)
    print (output.decode('ascii'))

ssh_client.close
```

4.3.4 程式分段說明

（1）首先匯入 paramiko、time 和 getpass 3 種模組。前兩種模組的用法已
經講過，這裡講解 getpass 模組。getpass 是 Python 的內建模組，無須透過
pip 下載安裝即可使用。它和 input() 函數一樣，都是 Python 的互動式功
能，用來提示使用者輸入密碼，區別是如果用 input() 輸入密碼，使用者輸
入的密碼是明文可見的，如果你身邊坐了其他人，密碼就這麼曝露了。而
透過 getpass 輸入的密碼則是隱藏不可見的，安全性很高，所以強烈建議
使用 getpass 來輸入密碼，使用 input() 來輸入使用者名稱。註：getpass 在
Windows 中有 bug，輸出的密碼依然明文可見，但是不影響指令稿的執行。

```
import Paramiko
import time
import getpass

username = input('Username: ')
password = getpass.getpass('Password: ')
```

（2）因為 5 個交換機 SW1 ～ SW5 的 IP 位址是連續的（192.168.2.11-
15），我們可以配合 for i in range(11，16) 做一個簡單的 for 循環來檢查
11 ～ 15 的反覆運算值（在 Python 3 中，range() 函數的傳回值不再是串
列，而是一組反覆運算值），然後以此配合下一行程式 ip = "192.168.2." +

str(i) 來實現循環（批次）登入交換機 SW1 ～ SW5。注意：這裡的 i 是整數，整數不能和字串做連接，所以要用 str(i) 先將 i 轉化成字串。

```
for i in range(11,16):
    ip = "192.168.2." + str(i)
    ssh_client = Paramiko.SSHClient()
    ssh_client.set_missing_host_key_policy(Paramiko.AutoAddPolicy())
    ssh_client.connect(hostname=ip，username=username，password=password)
    print ("Successfully connect to "，ip)
    command = ssh_client.invoke_shell()
    command.send("configure terminal\n")
```

（3）同樣的道理，我們要建立 VLAN 10 ～ VLAN 20，這些 VLAN Id 是連續的，所以又可以配合一個簡單的 for 循環 for n in range (10，21) 來達到循環設定 VLAN 10 ～ VLAN 20 的目的，這裡使用的是巢狀結構 for 循環，需要注意縮排。每建立一個 VLAN，都先列印內容 "print ("Creating VLAN " + str(n))" 來提示使用者目前正在建立的 VLAN。每個 VLAN 的命名格式都是 Python_VLAN XX，例如 VLAN 10 的名字是 Python_VLAN 10，VLAN 11 的名字是 Python_VLAN 11，依此類推。每建立一個 VLAN 之間都需要 1s 的間隔。

```
for n in range (10, 21):
print ("Creating VLAN " + str(n))
command.send("vlan " + str(n) +  "\n")
command.send("name Python_VLAN " + str(n) +  "\n")
time.sleep(1)
```

（4）最後儲存設定，間隔 2s 後列印出回應內容，並關閉 SSH。

```
command.send("end\n")
command.send("wr mem\n")
time.sleep(2)
output = command.recv(65535)
print (output)

ssh_client.close
```

4.3.5 驗證

（1）因列印出的回應內容過長，這裡只截取自動登入 SW1、SW2 和 SW3 做設定的部分程式。可以看到：當執行指令稿後系統提示輸入使用者名稱和密碼時，我們輸入的使用者名稱是可見的，密碼是不可見的，原因就是輸入使用者名稱時我們使用的是 input()，輸入密碼時我們用的是 getpass. getpass()，如下圖所示。

```
[root@CentOS-Python ~]# python3.8 lab1.py
Username: python
Password:
Successfully connect to  192.168.2.11
Creating VLAN 10
Creating VLAN 11
Creating VLAN 12
Creating VLAN 13
Creating VLAN 14
Creating VLAN 15
Creating VLAN 16
Creating VLAN 17
Creating VLAN 18
Creating VLAN 19
Creating VLAN 20


*******************************************************************
* IOSv is strictly limited to use for evaluation, demonstration and IOS  *
* education. IOSv is provided as-is and is not supported by Cisco's      *
* Technical Advisory Center. Any use or disclosure, in whole or in part, *
* of the IOSv Software or Documentation to any third party for any       *
* purposes is expressly prohibited except as otherwise authorized by     *
* Cisco in writing.                                                      *
*******************************************************SW1#configure terminal
Enter configuration commands, one per line.  End with CNTL/Z.
SW1(config)#vlan 10
SW1(config-vlan)#name Python_VLAN 10
SW1(config-vlan)#vlan 11
SW1(config-vlan)#name Python_VLAN 11
SW1(config-vlan)#vlan 12
SW1(config-vlan)#name Python_VLAN 12
SW1(config-vlan)#vlan 13
SW1(config-vlan)#name Python_VLAN 13
SW1(config-vlan)#vlan 14
SW1(config-vlan)#name Python_VLAN 14
SW1(config-vlan)#vlan 15
SW1(config-vlan)#name Python_VLAN 15
SW1(config-vlan)#vlan 16
SW1(config-vlan)#name Python_VLAN 16
SW1(config-vlan)#vlan 17
SW1(config-vlan)#name Python_VLAN 17
SW1(config-vlan)#
Successfully connect to  192.168.2.12
Creating VLAN 10
Creating VLAN 11
Creating VLAN 12
Creating VLAN 13
Creating VLAN 14
Creating VLAN 15
Creating VLAN 16
```

```
Creating VLAN 17
Creating VLAN 18
Creating VLAN 19
Creating VLAN 20

***********************************************************************
* IOSv is strictly limited to use for evaluation, demonstration and IOS  *
* education. IOSv is provided as-is and is not supported by Cisco's       *
* Technical Advisory Center. Any use or disclosure, in whole or in part,  *
* of the IOSv Software or Documentation to any third party for any        *
* purposes is expressly prohibited except as otherwise authorized by      *
* Cisco in writing.                                                       *
***********************************************************************
SW2#configure terminal
Enter configuration commands, one per line.  End with CNTL/Z.
SW2(config)#vlan 10
SW2(config-vlan)#name Python_VLAN 10
SW2(config-vlan)#vlan 11
SW2(config-vlan)#name Python_VLAN 11
SW2(config-vlan)#vlan 12
SW2(config-vlan)#name Python_VLAN 12
SW2(config-vlan)#vlan 13
SW2(config-vlan)#name Python_VLAN 13
SW2(config-vlan)#vlan 14

Successfully connect to  192.168.2.13
Creating VLAN 10
Creating VLAN 11
Creating VLAN 12
Creating VLAN 13
Creating VLAN 14
Creating VLAN 15
Creating VLAN 16
Creating VLAN 17
Creating VLAN 18
Creating VLAN 19
Creating VLAN 20

***********************************************************************
* IOSv is strictly limited to use for evaluation, demonstration and IOS  *
* education. IOSv is provided as-is and is not supported by Cisco's       *
* Technical Advisory Center. Any use or disclosure, in whole or in part,  *
* of the IOSv Software or Documentation to any third party for any        *
* purposes is expressly prohibited except as otherwise authorized by      *
* Cisco in writing.                                                       *
***********************************************************************
SW3#configure terminal
Enter configuration commands, one per line.  End with CNTL/Z.
SW3(config)#vlan 10
SW3(config-vlan)#name Python_VLAN 10
```

（2）依次登入 5 個交換機驗證設定。

執行程式後，SW1 的設定如下圖所示。

```
SW1#show vlan b

VLAN Name                             Status    Ports
---- -------------------------------- --------- -------------------------------
1    default                          active    Gi0/0, Gi0/1, Gi0/2, Gi0/3
                                                Gi1/0, Gi1/1, Gi1/2, Gi1/3
                                                Gi2/0, Gi2/1, Gi2/2, Gi2/3
                                                Gi3/0, Gi3/1, Gi3/2, Gi3/3
10   Python_VLAN 10                   active
11   Python_VLAN 11                   active
12   Python_VLAN 12                   active
13   Python_VLAN 13                   active
14   Python_VLAN 14                   active
15   Python_VLAN 15                   active
16   Python_VLAN 16                   active
17   Python_VLAN 17                   active
18   Python_VLAN 18                   active
19   Python_VLAN 19                   active
20   Python_VLAN 20                   active
1002 fddi-default                     act/unsup
1003 token-ring-default               act/unsup
1004 fddinet-default                  act/unsup
1005 trnet-default                    act/unsup
SW1#
```

執行程式後，SW2 的設定如下圖所示。

```
SW2#show vlan b

VLAN Name                             Status    Ports
---- -------------------------------- --------- -------------------------------
1    default                          active    Gi0/0, Gi0/1, Gi0/2, Gi0/3
                                                Gi1/0, Gi1/1, Gi1/2, Gi1/3
                                                Gi2/0, Gi2/1, Gi2/2, Gi2/3
                                                Gi3/0, Gi3/1, Gi3/2, Gi3/3
10   Python_VLAN 10                   active
11   Python_VLAN 11                   active
12   Python_VLAN 12                   active
13   Python_VLAN 13                   active
14   Python_VLAN 14                   active
15   Python_VLAN 15                   active
16   Python_VLAN 16                   active
17   Python_VLAN 17                   active
18   Python_VLAN 18                   active
19   Python_VLAN 19                   active
20   Python_VLAN 20                   active
1002 fddi-default                     act/unsup
1003 token-ring-default               act/unsup
1004 fddinet-default                  act/unsup
1005 trnet-default                    act/unsup
SW2#
```

執行程式後，SW3 的設定如下圖所示。

```
SW3#show vlan b

VLAN Name                             Status    Ports
---- -------------------------------- --------- -------------------------------
1    default                          active    Gi0/0, Gi0/1, Gi0/2, Gi0/3
                                                Gi1/0, Gi1/1, Gi1/2, Gi1/3
                                                Gi2/0, Gi2/1, Gi2/2, Gi2/3
                                                Gi3/0, Gi3/1, Gi3/2, Gi3/3
10   Python_VLAN 10                   active
11   Python_VLAN 11                   active
12   Python_VLAN 12                   active
13   Python_VLAN 13                   active
14   Python_VLAN 14                   active
15   Python_VLAN 15                   active
16   Python_VLAN 16                   active
17   Python_VLAN 17                   active
18   Python_VLAN 18                   active
19   Python_VLAN 19                   active
20   Python_VLAN 20                   active
1002 fddi-default                     act/unsup
1003 token-ring-default               act/unsup
1004 fddinet-default                  act/unsup
1005 trnet-default                    act/unsup
SW3#
```

執行程式後，SW4 的設定如下圖所示。

```
SW4#show vlan b

VLAN Name                             Status    Ports
---- -------------------------------- --------- -------------------------------
1    default                          active    Gi0/0, Gi0/1, Gi0/2, Gi0/3
                                                Gi1/0, Gi1/1, Gi1/2, Gi1/3
                                                Gi2/0, Gi2/1, Gi2/2, Gi2/3
                                                Gi3/0, Gi3/1, Gi3/2, Gi3/3
10   Python_VLAN 10                   active
11   Python_VLAN 11                   active
12   Python_VLAN 12                   active
13   Python_VLAN 13                   active
14   Python_VLAN 14                   active
15   Python_VLAN 15                   active
16   Python_VLAN 16                   active
17   Python_VLAN 17                   active
18   Python_VLAN 18                   active
19   Python_VLAN 19                   active
20   Python_VLAN 20                   active
1002 fddi-default                     act/unsup
1003 token-ring-default               act/unsup
1004 fddinet-default                  act/unsup
1005 trnet-default                    act/unsup
SW4#
```

執行程式後，SW5 的設定如下圖所示。

```
SW5#show vlan b

VLAN Name                          Status    Ports
---- ------------------------------ --------- -------------------------------
1    default                        active    Gi0/0, Gi0/1, Gi0/2, Gi0/3
                                              Gi1/0, Gi1/1, Gi1/2, Gi1/3
                                              Gi2/0, Gi2/1, Gi2/2, Gi2/3
                                              Gi3/0, Gi3/1, Gi3/2, Gi3/3
10   Python_VLAN 10                 active
11   Python_VLAN 11                 active
12   Python_VLAN 12                 active
13   Python_VLAN 13                 active
14   Python_VLAN 14                 active
15   Python_VLAN 15                 active
16   Python_VLAN 16                 active
17   Python_VLAN 17                 active
18   Python_VLAN 18                 active
19   Python_VLAN 19                 active
20   Python_VLAN 20                 active
1002 fddi-default                   act/unsup
1003 token-ring-default             act/unsup
1004 fddinet-default                act/unsup
1005 trnet-default                  act/unsup
SW5#
```

4.4 實驗 2：批次登入不同網段的交換機

在生產環境中，交換機的管理 IP 位址基本不可能像實驗 1 中那樣是連續的，有些交換機的管理 IP 位址甚至在不同的網段。在這種情況下，我們就不能簡單地用 for 循環來登入交換機了。我們要額外建立一個文字檔，把需要登入的交換機的管理 IP 位址全部寫進去，然後用 for 循環配合 open() 函數來讀取該文件中的管理 IP 位址，進一步達到批次登入交換機的目的。

4.4.1 實驗目的

- 透過 Python 指令稿批次登入所有交換機，並在每個交換機上都開啟 EIGRP。

4.4.2 實驗準備

（1）把 SW5 的管理位址從 192.168.2.15 改成 192.168.2.55。

```
S5#conf t
Enter configuration commands, one per line.  End with CNTL/Z.
S5(config)#int vlan 1
S5(config-if)#ip add 192.168.2.55 255.255.255.0
S5(config-if)#end
S5#
*Mar 26 07:03:04.886: %SYS-5-CONFIG_I: Configured from console by console
S5#
```

（2）在 CentOS 上建立一個名為 ip_list 的 TXT 檔案，把所有交換機的管理 IP 位址都放進去，注意該檔案和等下要建立的指令稿位於同一個資料夾下，如下圖所示。

```
[root@CentOS-Python ~]# cat ip_list.txt
192.168.2.11
192.168.2.12
192.168.2.13
192.168.2.14
192.168.2.55
[root@CentOS-Python ~]#
```

（3）在執行程式前，檢查 5 台交換機的設定，確認它們都沒有開啟 EIGRP。

執行程式前，SW1 的設定如下圖所示。

```
SW1#show run | s router eigrp
SW1#
```

執行程式前，SW2 的設定如下圖所示。

```
SW2#show run | s router eigrp
SW2#
```

執行程式前，SW3 的設定如下圖所示。

```
SW3#show run | s router eigrp
SW3#
```

執行程式前，SW4 的設定如下圖所示。

```
SW4#show run | s router eigrp
SW4#
```

執行程式前，SW5 的設定如下圖所示。

```
SW5#show run | s router eigrp
SW5#
```

（4）在主機上建立實驗 2 的指令稿，將其命名為 lab2.py，如下圖所示。

```
[root@CentOS-Python ~]# vi lab2.py
```

4.4.3 實驗程式

將下列程式寫入指令稿 lab2.py。

```
import paramiko
import time
from getpass import getpass

username = input('Username: ')
password = getpass('password: ')

f = open("ip_list.txt", "r")
for line in f.readlines():
    ip = line.strip()
```

```
ssh_client = paramiko.SSHClient()
ssh_client.set_missing_host_key_policy(paramiko.AutoAddPolicy())
ssh_client.connect(hostname=ip, username=username, password=password)
print ("Successfully connect to ", ip)
remote_connection = ssh_client.invoke_shell()
remote_connection.send("conf t\n")
remote_connection.send("router eigrp 1\n")
remote_connection.send("end\n")
remote_connection.send("wr mem\n")
time.sleep(1)
output = remote_connection.recv(65535)
print (output.decode("ascii"))


f.close()
ssh_client.close
```

4.4.4 程式分段說明

（1）和實驗 1 稍有不同，我們在匯入 getpass 模組時用的是 from getpass import getpass，因此我們可以把 getpass.getpass('password: ') 簡寫成 getpass('password: ')。

```
import paramiko
import time
from getpass import getpass


username = input('Username: ')
password = getpass('password: ')
```

（2）用 open() 函數開啟之前建立好的包含 5 個交換機的管理 IP 位址的文件（ip_list.txt），透過 for 循環來依次檢查 readlines() 方法傳回的串列中的每個元素（即每個交換機的管理 IP 位址），即可達到批次依次登入 SW1 ～ SW5 的目的，即使這 5 個交換機的管理 IP 位址不是連續的。

```
f = open("ip_list.txt", "r")
for line in f.readlines():
    ip = line.strip()
    ssh_client = paramiko.SSHClient()
    ssh_client.set_missing_host_key_policy(paramiko.AutoAddPolicy())
    ssh_client.connect(hostname=ip，username=username，password=password)
    print ("Successfully connect to ", ip)
```

（3）登入每台交換機後設定 EIGRP，將回應內容列印出來。

```
remote_connection = ssh_client.invoke_shell()
remote_connection.send("conf t\n")
remote_connection.send("router eigrp 1\n")
remote_connection.send("end\n")
remote_connection.send("wr mem\n")
time.sleep(1)
output = remote_connection.recv(65535)
print (output.decode("ascii"))
```

（4）檔案有開有關，指令稿結束前用 close() 關掉 ip_list.txt 文件，並且關閉 SSH 連結。

```
f.close()
ssh_client.close
```

4.4.5 驗證

（1）因列印出的回應內容過長，這裡只截取自動登入 SW1、SW2 做設定的部分程式，可以看見程式中自動登入了每個交換機開啟 EIGRP 並儲存設定，隨後退出，如下圖所示。

```
[root@CentOS-Python ~]# python3.8 lab2.py
Username: python
password:
Successfully connect to  192.168.2.11

********************************************************************
* IOSv is strictly limited to use for evaluation, demonstration and IOS  *
* education. IOSv is provided as-is and is not supported by Cisco's     *
* Technical Advisory Center. Any use or disclosure, in whole or in part, *
* of the IOSv Software or Documentation to any third party for any      *
* purposes is expressly prohibited except as otherwise authorized by    *
* Cisco in writing.                                                     *
********************************************************************
SW1#conf t
Enter configuration commands, one per line.  End with CNTL/Z.
SW1(config)#router eigrp 1
SW1(config-router)#end
SW1#wr mem
Building configuration...

Successfully connect to  192.168.2.12

********************************************************************
* IOSv is strictly limited to use for evaluation, demonstration and IOS  *
* education. IOSv is provided as-is and is not supported by Cisco's     *
* Technical Advisory Center. Any use or disclosure, in whole or in part, *
* of the IOSv Software or Documentation to any third party for any      *
* purposes is expressly prohibited except as otherwise authorized by    *
* Cisco in writing.                                                     *
********************************************************************
SW2#conf t
Enter configuration commands, one per line.  End with CNTL/Z.
SW2(config)#router eigrp 1
SW2(config-router)#end
SW2#wr mem
Building configuration...
```

（2）依次登入 5 個交換機驗證設定。

執行程式後，SW1 的設定如下圖所示。

```
SW1#show run | s router eigrp
router eigrp 1
SW1#
```

執行程式後，SW2 的設定如下圖所示。

```
SW2#show run | s router eigrp
router eigrp 1
SW2#
```

執行程式後，SW3 的設定如下圖所示。

```
SW3#show run | s router eigrp
router eigrp 1
SW3#
```

執行程式後，SW4 的設定如下圖所示。

```
SW4#show run | s router eigrp
router eigrp 1
SW4#
```

執行程式後，SW5 的設定如下圖所示。

```
SW5#show run | s router eigrp
router eigrp 1
SW5#
```

4.5 實驗 3：異常處理的應用

在網路裝置數量超過千台甚至上萬台的大型企業網中，難免會遇到某些裝置的管理 IP 位址不通、SSH 連接失敗的情況，裝置數量越多，這種情況發生的機率越高。這個時候如果你想用 Python 批次設定所有的裝置，就一定要注意這種情況，很可能你的指令稿執行了還不到一半就因為中間某

一個連接不通的裝置而停止了。例如你有 5000 台交換機需要統一更改本
機使用者名稱和密碼，前 500 台交換機的連通性都沒有問題，第 501 台
交換機因為某個網路問題導致管理 IP 位址無法連接，SSH 連不上，此時
Python 會傳回一個 "socket.error: [Errno 10060] A connection attempt failed
because the connected party did not properly respond after a period of time，
or established connection failed because connected host has failed to respond"
錯誤，然後指令稿就此停住！指令稿不會再對剩下的 4500 台交換機做設
定，也就表示「掛機」失敗！如下圖所示。

```
Traceback (most recent call last):
  File "Verify configuration violation.py", line 23, in <module>
    ssh_client.connect(hostname=ip,username=username,password=password,look_for_
keys=False)
  File "C:\Python27\lib\site-packages\paramiko\client.py", line 338, in connect
    retry_on_signal(lambda: sock.connect(addr))
  File "C:\Python27\lib\site-packages\paramiko\util.py", line 279, in retry_on_s
ignal
    return function()
  File "C:\Python27\lib\site-packages\paramiko\client.py", line 338, in <lambda>
    retry_on_signal(lambda: sock.connect(addr))
  File "C:\Python27\lib\socket.py", line 228, in meth
    return getattr(self._sock,name)(*args)
socket.error: [Errno 10060] A connection attempt failed because the connected pa
rty did not properly respond after a period of time, or established connection f
ailed because connected host has failed to respond
```

同樣的問題也會發生在當你輸入了錯誤的交換機使用者名稱和密碼時，
或某些交換機和其他大部分交換機使用者名稱和密碼不一致時（因為我
們只能輸入一次使用者名稱和密碼，使用者名稱和密碼不一致會導致無
法登入個別交換機的情況發生），也許你會問大型企業網不都是統一設定
AAA 配合 TACACS 或 RADIUS 做使用者存取管理嗎？怎麼還會出現登
入帳號和密碼不一致的問題？這個現象就發生在筆者目前所任職的沙烏地
阿拉伯阿布都拉國王科技大學，學校裡的 TACACS 伺服器（思科 ACS）
已經服役 9 年，目前的問題是每天早晨 8 點左右該 ACS 會「故障」，必須
手動重新啟動 ACS 所在的伺服器才能解決問題，在 ACS 無法正常執行期

間，我們只能透過網路裝置的本機帳號和密碼登入。鑑於此，我們已經部署了思科的 ISE 來替代 ACS 做 TACACS 伺服器，但由於學校網路過於龐大，遷徙過程漫長，就導致了部分裝置已經遷徙，使用了 ISE 設定的帳號和密碼；而另一部分還沒有遷徙的裝置在 ACS 出問題的時候只能用本機的帳號和密碼，這就出現了兩套帳號和密碼的情況，後果就是使用 Paramiko 來 SSH 登入網路裝置的 Python 會傳回 "Paramiko.ssh_exception. AuthenticationException: Authentication failed" 的錯誤，如下圖所示，導致指令稿戛然而止，無法繼續執行。

```
Traceback (most recent call last):
  File "Verify configuration violation.py", line 25, in <module>
    ssh_client.connect(hostname=ip,username=username,password=password,look_for_
keys=False)
  File "C:\Python27\lib\site-packages\paramiko\client.py", line 424, in connect
    passphrase.
  File "C:\Python27\lib\site-packages\paramiko\client.py", line 714, in _auth
    raise saved_exception
paramiko.ssh_exception.AuthenticationException: Authentication failed.
```

解決上述兩個問題的方法也許你已經想到了，就是使用我們在第 3 章中講到的異常處理。下面我們就用實驗來示範異常處理在上述兩種情況中的應用。

4.5.1 實驗目的

- 使用實驗 1 和實驗 2 中的網路拓撲，將交換機 SW3（192.168.2.13）使用者名稱 python 的密碼從 123 改為 456，並將 SW4（192.168.2.14）的 Gi0/0 通訊埠斷掉。

- 建立一個帶有 try…except… 異常處理敘述的指令稿來批次在交換機 SW1 ～ SW5 上執行 show clock 指令，讓指令稿在 SW3、SW4 分別因為使用者名稱和密碼不符合，以及連通性出現故障的情況下，依然可以不受干擾，進而完成剩餘的設定。

4.5.2 實驗準備

（1）首先將 SW3 使用者名稱 python 的密碼從 123 改為 456，如下圖所示。

```
SW3#conf t
Enter configuration commands, one per line.  End with CNTL/Z.
SW3(config)#username python password 456
SW3(config)#end
```

（2）然後將 SW4 的通訊埠 Gi0/0 關掉，如下圖所示。

```
SW4#conf t
Enter configuration commands, one per line.  End with CNTL/Z.
SW4(config)#int gi0/0
SW4(config-if)#shutdown
SW4(config-if)#end
SW4#
*May  4 08:31:15.664: %SYS-5-CONFIG_I: Configured from console by console
SW4#
```

（3）在主機上建立一個名為 ip_list.txt 的文字檔，內含 SW1 ～ SW5 的管理 IP 位址。注意，在實驗 2 中，我們已經將 SW5 的管理 IP 位址從192.168.2.15 改成了 192.168.2.55，如下圖所示。

```
[root@CentOS-Python ~]# cat ip_list.txt
192.168.2.11
192.168.2.12
192.168.2.13
192.168.2.14
192.168.2.55
[root@CentOS-Python ~]#
```

（4）延續實驗 2 的想法，我們在主機上建立一個名為 cmd.txt 的文字檔，寫入我們要在 SW1 ～ SW5 上執行的指令：show clock，如下圖所示。

```
[root@CentOS-Python ~]#
[root@CentOS-Python ~]# cat cmd.txt
show clock
[root@CentOS-Python ~]#
```

（5）最後建立實驗 3 的指令檔 lab3.py，如下圖所示。

```
[root@CentOS-Python ~]#
[root@CentOS-Python ~]# vi lab3.py
```

4.5.3 實驗程式

將下列程式寫入指令稿 lab3.py。

```python
import paramiko
import time
import getpass
import sys
import socket

username = input('Username: ')
password = getpass.getpass('password: ')
ip_file = sys.argv[1]
cmd_file = sys.argv[2]

switch_with_authentication_issue = []
switch_not_reachable = []

iplist = open(ip_file, 'r')
for line in iplist.readlines():
    try:
        ip = line.strip()
        ssh_client = paramiko.SSHClient()
        ssh_client.set_missing_host_key_policy(paramiko.AutoAddPolicy())
        ssh_client.connect(hostname=ip, username=username, password=password)
        print ("You have successfully connect to ", ip)
        command = ssh_client.invoke_shell()
        cmdlist = open(cmd_file, 'r')
        cmdlist.seek(0)
        for line in cmdlist.readlines():
```

```
        command.send(line + "\n")
        time.sleep(2)
        cmdlist.close()
        output = command.recv(65535)
        print (output.decode( 'ascii' ))
    except paramiko.ssh_exception.AuthenticationException:
        print ("User authentication failed for " + ip + ".")
        switch_with_authentication_issue.append(ip)
    except socket.error:
        print (ip +  " is not reachable.")
        switch_not_reachable.append(ip)

iplist.close()
ssh_client.close

print ('\nUser authentication failed for below switches: ')
for i in switch_with_authentication_issue:
    print (i)

print ('\nBelow switches are not reachable: ')
for i in switch_not_reachable:
    print (i)
```

4.5.4 程式分段說明

（1）為了使用異常處理來應對網路裝置無法連接引起的 socket.error，必須
匯入 Python 的內建模組 socket。

```
import paramiko
import time
import getpass
import sys
import socket

username = input('Username: ')
```

```
password = getpass.getpass('password: ')
ip_file = sys.argv[1]
cmd_file = sys.argv[2]
```

（2）建立兩個空串列，分別命名為 switch_with_authentication_issue 和
switch_not_ reachable，它們的作用是在指令稿最後配合 for 循環來統計哪
些裝置是因為認證問題而無法登入的，哪些裝置是因為裝置本身無法連接
而無法登入的。

```
switch_with_authentication_issue = []
switch_not_reachable = []
```

（3）在 for 循環下使用 try…except… 異常處理敘述。當 SSH 登入交換機
時，如果使用者名稱和密碼不正確，Python 會顯示出錯 "Paramiko.ssh_
exception.AuthenticationException"，因此我們用 except Paramiko.ssh_
exception.AuthenticationException: 來應對該異常，一旦有交換機出現使用
者名稱和密碼不正確的情況，列印出 "User authentication failed for [交換
機 IP]" 來取代前面的 "Paramiko.ssh_exception.AuthenticationException" 錯
誤訊息，然後將出現該異常的交換機的管理 IP 位址用串列的 append() 方
法放入 switch_with_authentication_issue 串列中。同理，用 except socket.
error: 來應對交換機無法連接時傳回的錯誤 "socket.error: [Errno 10060] A
connection attempt failed because the connected party did not properly respond
after a period of time, or established connection failed because connected host
has failed to respond"，並列印出 "[交換機 IP] is not reachable" 來取代上述
錯誤訊息，然後將出現該錯誤的交換機的管理 IP 位址用串列的 append()
方法放入 switch_not_reachable 串列中。

```
iplist = open(ip_file, 'r')
for line in iplist.readlines():
    try:
        ip = line.strip()
        ssh_client = paramiko.SSHClient()
```

```
        ssh_client.set_missing_host_key_policy(paramiko.AutoAddPolicy())
        ssh_client.connect(hostname=ip，username=username，password=password)
        print ("You have successfully connect to ", ip)
        command = ssh_client.invoke_shell()
        cmdlist = open(cmd_file, 'r')
        cmdlist.seek(0)
        for line in cmdlist.readlines():
            command.send(line + "\n")
        time.sleep(2)
        cmdlist.close()
        output = command.recv(65535)
        print (output.decode( 'ascii' ))
    except paramiko.ssh_exception.AuthenticationException:
        print ("User authentication failed for " + ip + ".")
        switch_with_authentication_issue.append(ip)
    except socket.error:
        print (ip +  " is not reachable.")
        switch_not_reachable.append(ip)

iplist.close()
ssh_client.close
```

（4）最後使用 for 循環，列印出 switch_with_authentication_issue 和 switch_
not_reachable 兩個串列中的元素，這樣就能清楚看到有哪些交換機的使用
者名稱和密碼驗證失敗，哪些交換機的管理 IP 位址無法連接。

```
print ('\nUser authentication failed for below switches: ')
for i in switch_with_authentication_issue:
    print (i)

print ('\nBelow switches are not reachable: ')
for i in switch_not_reachable:
    print (i)
```

4.5.5 驗證

如下圖所示，重點部分已經用線標記。

```
[root@localhost ~]# python lab4.py ip_list.txt cmd.txt
Username: python
password:
You have successfully connect to  192.168.2.11

**************************************************************************
* IOSv - Cisco Systems Confidential                                      *
*                                                                        *
* This software is provided as is without warranty for internal         *
* development and testing purposes only under the terms of the Cisco     *
* Early Field Trial agreement.  Under no circumstances may this software *
* be used for production purposes or deployed in a production            *
* environment.                                                           *
*                                                                        *
* By using the software, you agree to abide by the terms and conditions  *
* of the Cisco Early Field Trial Agreement as well as the terms and      *
* conditions of the Cisco End User License Agreement at                  *
* http://www.cisco.com/go/eula                                           *
*                                                                        *
* Unauthorized use or distribution of this software is expressly         *
* Prohibited.                                                            *
**************************************************************************
```

```
S1#show clock
*06:07:20.072 PDT Sat Mar 30 2019
S1#
S1#
You have successfully connect to  192.168.2.12

**************************************************************************
* IOSv - Cisco Systems Confidential                                      *
*                                                                        *
* This software is provided as is without warranty for internal         *
* development and testing purposes only under the terms of the Cisco     *
* Early Field Trial agreement.  Under no circumstances may this software *
* be used for production purposes or deployed in a production            *
* environment.                                                           *
*                                                                        *
* By using the software, you agree to abide by the terms and conditions  *
* of the Cisco Early Field Trial Agreement as well as the terms and      *
* conditions of the Cisco End User License Agreement at                  *
* http://www.cisco.com/go/eula                                           *
*                                                                        *
* Unauthorized use or distribution of this software is expressly         *
* Prohibited.                                                            *
**************************************************************************
```

```
S2#show clock
*13:06:52.531 UTC Sat Mar 30 2019
S2#
S2#
User authentication failed for 192.168.2.13.
192.168.2.14 is not reacnaole.
You nave successfully-connect to  192.168.2.55

*********************************************************************
* IOSv - Cisco Systems Confidential                                 *
*                                                                   *
* This software is provided as is without warranty for internal     *
* development and testing purposes only under the terms of the Cisco *
* Early Field Trial agreement.  Under no circumstances may this software *
* be used for production purposes or deployed in a production       *
* environment.                                                      *
*                                                                   *
* By using the software, you agree to abide by the terms and conditions *
* of the Cisco Early Field Trial Agreement as well as the terms and *
* conditions of the Cisco End User License Agreement at             *
* http://www.cisco.com/go/eula                                      *
*                                                                   *
* Unauthorized use or distribution of this software is expressly    *
* Prohibited.                                                       *
*********************************************************************
S5#show clock
*13:08:41.053 UTC Sat Mar 30 2019
S5#
S5#
User authentication failed for below switches:
192.168.2.13

Below switches are not reachable:
192.168.2.14
```

（1）本應出現的 "Paramiko.ssh_exception.AuthenticationException" 錯誤已經被 "User authentication failed for 192.168.2.13" 取代，並且 Python 指令稿並未就此停止執行，而是繼續嘗試登入下一個交換機 SW4，也就是 192.168.2.14。

（2）本應出現的 "socket.error: [Errno 10060] A connection attempt failed because the connected party did not properly respond after a period of time，or established connection failed because connected host has failed to respond" 錯誤已經被 "192.168.2.14 is not reachable" 取代，並且 Python 指令稿並未就此停止執行，而是繼續嘗試登入下一個交換機 SW5，也就是 192.168.2.55。

（3）在指令稿的最後，可以看到哪些交換機出現了使用者名稱和密碼認證失敗的情況，哪些交換機出現了管理 IP 位址無法連接的情況。

4.6 實驗 4：用**Python**實現網路裝置的設定備份

將網路裝置的設定做備份是網路運行維護中必不可少的一項工作，根據公司的規模和要求不同，管理層可能會要求對網路裝置的設定做月備、周備甚至日備。傳統的備份思科交換機設定的辦法是手動 SSH 遠端登入裝置，然後輸入指令 term len 0 和 show run，將回應內容手動複製、貼上到一個 TXT 或 Word 文字檔中，效率十分不佳，在有成百上千台裝置需要備份的網路中尤為明顯。實驗 4 將示範如何用 Python 來解決這個困擾了傳統網路工程師很多年的網路運行維護痛點問題。

4.6.1 實驗目的

- 在 CentOS 8 主機上開啟 FTP Server 服務（鑑於 TFTP 在 CentOS 8 上安裝使用的複雜程度，這裡用 FTP 替代），建立 Python 指令稿，將 SW1 ～ SW5 的 running configuration 備份儲存到 TFTP 伺服器上。

4.6.2 實驗準備

（1）將 SW5 的管理 IP 位址改回 192.168.2.15，將 SW3 的使用者名稱 python 的密碼從 456 改回 123，將 SW4 的 Gi0/0 通訊埠重新開啟，如下圖所示。

```
SW5#
SW5#conf t
Enter configuration commands, one per line.  End with CNTL/Z.
SW5(config)#int vlan 1
SW5(config-if)#ip add 192.168.2.15 255.255.255.0
SW5(config-if)#end
SW5#wr mem
```

```
SW3#
SW3#conf t
Enter configuration commands, one per line.  End with CNTL/Z.
SW3(config)#username python password 123
SW3(config)#end
SW3#wr mem
Building configuration...
```

```
SW4#conf t
Enter configuration commands, one per line.  End with CNTL/Z.
SW4(config)#int gi0/0
SW4(config-if)#no shut
SW4(config-if)#end
SW4#
*May  8 10:05:55.505: %SYS-5-CONFIG_I: Configured from console by console
SW4#
*May  8 10:05:56.060: %LINK-3-UPDOWN: Interface GigabitEthernet0/0, changed state to up
*May  8 10:05:57.064: %LINEPROTO-5-UPDOWN: Line protocol on Interface GigabitEthernet0/0, changed state to up
SW4#
```

（2）將 CentOS 上的 ip_list.txt 裡 SW5 的管理 IP 位址也改回 192.168.2.15，
如下圖所示。

```
[root@CentOS-Python ~]# cat ip_list.txt
192.168.2.11
192.168.2.12
192.168.2.13
192.168.2.14
192.168.2.15
[root@CentOS-Python ~]#
```

（3）在 CentOS 8 主機上輸入下列指令下載安裝 vsftpd（FTP 服務），安裝
前需要確認主機是否可連通外網。

```
dnf install vsftpd -y
```

（4）安裝完成後，在 CentOS 8 主機上輸入下面兩個指令分別讓 CentOS 在
目前和開機時啟動 vsftpd 服務，如下圖所示。

```
[root@CentOS-Python ~]# dnf install vsftpd -y
Last metadata expiration check: 0:06:21 ago on Fri 08 May 2020 04:22:58 AM EDT.
Dependencies resolved.
================================================================================
 Package                                        Architecture
================================================================================
Installing:
 vsftpd                                         x86_64
 
Transaction Summary
================================================================================
Install  1 Package

Total download size: 180 k
Installed size: 359 k
Downloading Packages:
vsftpd-3.0.3-28.el8.x86_64.rpm
--------------------------------------------------------------------------------
Total
Running transaction check
Transaction check succeeded.
Running transaction test
Transaction test succeeded.
Running transaction
  Preparing        :
  Installing       : vsftpd-3.0.3-28.el8.x86_64
  Running scriptlet: vsftpd-3.0.3-28.el8.x86_64
  Verifying        : vsftpd-3.0.3-28.el8.x86_64

Installed:
  vsftpd-3.0.3-28.el8.x86_64

Complete!
[root@CentOS-Python ~]#
```

```
systemctl start vsftpd
systemctl enable vsftpd
```

```
[root@CentOS-Python ~]# systemctl start vsftpd
[root@CentOS-Python ~]# systemctl enable vsftpd
Created symlink /etc/systemd/system/multi-user.target.wants/vsftpd.service → /usr/lib/s
ystemd/system/vsftpd.service.
[root@CentOS-Python ~]#
```

（5）輸入下列指令確認 vsfptd 已經被啟動執行。

```
systemctl status vsftpd
```

```
[root@CentOS-Python ~]# systemctl status vsftpd
● vsftpd.service - Vsftpd ftp daemon
   Loaded: loaded (/usr/lib/systemd/system/vsftpd.service; enabled; vendor preset: disabled)
   Active: active (running) since Fri 2020-05-08 04:35:25 EDT; 2min 58s ago
 Main PID: 33940 (vsftpd)
    Tasks: 1 (limit: 11343)
   Memory: 552.0K
   CGroup: /system.slice/vsftpd.service
           └─33940 /usr/sbin/vsftpd /etc/vsftpd/vsftpd.conf

May 08 04:35:25 CentOS-Python systemd[1]: Starting Vsftpd ftp daemon...
May 08 04:35:25 CentOS-Python systemd[1]: Started Vsftpd ftp daemon.
[root@CentOS-Python ~]#
```

（6）輸入下列指令關閉 CentOS 8 的防火牆（僅用作實驗示範，生產環境中建議修改防火牆策略）並驗證，如下圖所示。

```
systemctl stop firewalld
systemctl status firewalld
```

```
[root@CentOS-Python ~]# systemctl stop firewalld
[root@CentOS-Python ~]# systemctl status firewalld
● firewalld.service - firewalld - dynamic firewall daemon
   Loaded: loaded (/usr/lib/systemd/system/firewalld.service; enabled; vendor preset: enabled)
   Active: inactive (dead) since Fri 2020-05-08 05:12:23 EDT; 11s ago
     Docs: man:firewalld(1)
  Process: 1110 ExecStart=/usr/sbin/firewalld --nofork --nopid $FIREWALLD_ARGS (code=exited, status=0/SUCCESS)
 Main PID: 1110 (code=exited, status=0/SUCCESS)

May 08 04:22:20 CentOS-Python systemd[1]: Starting firewalld - dynamic firewall daemon...
May 08 04:22:20 CentOS-Python systemd[1]: Started firewalld - dynamic firewall daemon.
May 08 04:22:20 CentOS-Python firewalld[1110]: WARNING: AllowZoneDrifting is enabled. This is considered an insec
May 08 05:12:22 CentOS-Python systemd[1]: Stopping firewalld - dynamic firewall daemon...
May 08 05:12:23 CentOS-Python systemd[1]: Stopped firewalld - dynamic firewall daemon.
[root@CentOS-Python ~]#
```

（7）在 CentOS 上輸入下列指令分別建立新的使用者名稱 python，並根據提示設定密碼 python，使用者名稱 python 將稍後作為 FTP 的使用者使用名，而交換機的 running configuration 也將被儲存在 /home/python 資料夾下，如下圖所示。

```
useradd -create-home python
passwd python
```

```
[root@CentOS-Python ~]# useradd --create-home python
[root@CentOS-Python ~]# passwd python
Changing password for user python.
New password:
BAD PASSWORD: The password is shorter than 8 characters
Retype new password:
passwd: all authentication tokens updated successfully.
[root@CentOS-Python ~]#
```

（8）在主機上建立實驗 4 的指令稿，將其命名為 lab4.py。

```
[root@CentOS-Python ~]#
[root@CentOS-Python ~]# vi lab4.py
```

4.6.3 實驗程式

將下列程式寫入指令稿 lab4.py。

```python
import paramiko
import time
import getpass

username = input('Username: ')
password = getpass.getpass('password: ')

f = open("ip_list.txt")

for line in f.readlines():
    ip_address = line.strip()
    ssh_client = paramiko.SSHClient()
    ssh_client.set_missing_host_key_policy(paramiko.AutoAddPolicy())
    ssh_client.connect(hostname=ip_address,username=username,password=password)
    print ("Successfully connect to ", ip_address)
    command = ssh_client.invoke_shell()
    command.send("configure terminal\n")
    command.send("ip ftp username python\n")
    command.send("ip ftp password python\n")
    command.send("file prompt quiet\n")
    command.send("end\n")
    command.send("copy running-config ftp://192.168.2.1\n")
    time.sleep(5)
    output = command.recv(65535)
    print (output.decode('ascii'))

f.close()
ssh_client.close
```

4.6.4 程式分段說明

（1）實驗 4 的程式難度不大，下面這段程式的作用在實驗 2 中已經有詳細解釋，這裡不再贅述。

```
import paramiko
import time
import getpass

username = input('Username: ')
password = getpass.getpass('password: ')

f = open("ip_list.txt")

for line in f.readlines():
    ip_address = line.strip()
    ssh_client = paramiko.SSHClient()
    ssh_client.set_missing_host_key_policy(paramiko.AutoAddPolicy())
    ssh_client.connect(hostname=ip_address,username=username,password=password)
    print ("Successfully connect to ", ip_address)
    command = ssh_client.invoke_shell()
```

（2）首先我們在每個交換機中透過指令 ip ftp username python 和 ip ftp password python 建立 FTP 使用者名稱和密碼，該使用者名稱和密碼同我們在 CentOS 主機上建立的一樣。

```
command.send("configure terminal\n")
command.send("ip ftp username python\n")
command.send("ip ftp password python\n")
```

（3）開啟 file prompt quiet，然後將交換機的設定檔備份到 CentOS 主機。

```
command.send("file prompt quiet\n")
    command.send("end\n")
    command.send("copy running-config ftp://192.168.2.1\n")
```

指令 file prompt 用來修改交換機檔案操作的提醒方式有 alert、noisy 和 quiet 3 種模式，預設是 alert。該模式和 noisy 都會在使用者進行檔案操作時提示使用者確認目標主機位址及目的檔案名稱等參數，例如我們在交換機裡輸入指令 copy running-config ftp://192.168.2.1 將交換機的設定檔透過 FTP 備份到 CentOS 主機 192.168.2.1 後，如果使用 file prompt alert 或 file prompt noisy，則交換機都會提醒你對目標主機位址及目的檔案名稱進行確認，舉例如下。

```
SW1#conf t
Enter configuration commands, one per line.  End with CNTL/Z.
SW1(config)#file pr
SW1(config)#file prompt ?
  alert  Prompt only for destructive file operations
  noisy  Confirm all file operation parameters
  quiet  Seldom prompt for file operations
  <cr>

SW1(config)#file prompt alert
SW1(config)#end
SW1#
SW1#
SW1#
SW1#
*May  8 10:40:30.010: %SYS-5-CONFIG_I: Configured from console by console
SW1#copy running-config ftp://192.168.2.1
Address or name of remote host [192.168.2.1]?
Destination filename [sw1-confg]?
Writing swi-confg !
3822 bytes copied in 25.809 secs (148 bytes/sec)
SW1#
```

系統已經自動設定好目的檔案名稱，其格式為「交換機的 hostname-config」，如果你對這個系統預設設定好的目的檔案名稱沒有問題，那麼 alert 和 noisy 這兩種 file prompt 模式不但對我們沒有任何幫助，反而還會影響指令稿的執行，因此我們使用 quiet 模式。關於 quiet 模式的效果舉例如下。

```
SW1#conf t
Enter configuration commands, one per line.  End with CNTL/Z.
SW1(config)#file prompt quiet
SW1(config)#do copy running-config ftp://192.168.2.1
Writing sw1-confg !
3840 bytes copied in 25.533 secs (150 bytes/sec)
SW1(config)#
```

（4）後面部分的程式都是實驗 1 和 2 中講過的，不再贅述。

```
    time.sleep(3)
    output = command.recv(65535)
    print (output.decode('ascii'))


f.close()
```

4.6.5 驗證

（1）執行程式前，在 CentOS 主機上確認 /home/python 資料夾下沒有任何
檔案，如下圖所示。

```
[root@CentOS-Python ~]# cd /home/python/
[root@CentOS-Python python]# ls
[root@CentOS-Python python]#
```

（2）回到指令稿 lab4.py 所在的資料夾，執行指令稿 lab4.py 後，輸入交
換機的 SSH 使用者名稱和密碼然後看效果，這裡只截取指令稿在 SW1 和
SW2 上執行後的回應內容，如下圖所示。

```
[root@CentOS-Python python]# cd
[root@CentOS-Python ~]# python3.8 lab4.py
Username: python
password:
Successfully connect to  192.168.2.11

************************************************************
* IOSv is strictly limited to use for evaluation, demonstration and IOS  *
* education. IOSv is provided as-is and is not supported by Cisco's       *
* Technical Advisory Center. Any use or disclosure, in whole or in part,  *
* of the IOSv Software or Documentation to any third party for any        *
* purposes is expressly prohibited except as otherwise authorized by      *
* Cisco in writing.                                                       *
************************************************************
SW1#configure terminal
Enter configuration commands, one per line.  End with CNTL/Z.
SW1(config)#ip ftp username python
SW1(config)#ip ftp password python
SW1(config)#file prompt quiet
SW1(config)#end
SW1#copy running-config ftp://192.168.2.1
```

```
Successfully connect to  192.168.2.12

*************************************************************
* IOSv is strictly limited to use for evaluation, demonstration and IOS  *
* education. IOSv is provided as-is and is not supported by Cisco's       *
* Technical Advisory Center. Any use or disclosure, in whole or in part,  *
* of the IOSv Software or Documentation to any third party for any        *
* purposes is expressly prohibited except as otherwise authorized by      *
* Cisco in writing.                                                       *
*************************************************************
SW2#configure terminal
Enter configuration commands, one per line.  End with CNTL/Z.
SW2(config)#ip ftp username python
SW2(config)#ip ftp password python
SW2(config)#file prompt quiet
SW2(config)#end
SW2#copy running-config ftp://192.168.2.1

Successfully connect to  192.168.2.13

*************************************************************
* IOSv is strictly limited to use for evaluation, demonstration and IOS  *
* education. IOSv is provided as-is and is not supported by Cisco's       *
* Technical Advisory Center. Any use or disclosure, in whole or in part,  *
* of the IOSv Software or Documentation to any third party for any        *
* purposes is expressly prohibited except as otherwise authorized by      *
```

（3）指令稿執行完畢後，回到 /home/python，此時可以看到 SW1 ～ SW5 的 running config 都被成功備份到該資料夾下，如下圖所示。

```
[root@CentOS-Python ~]#
[root@CentOS-Python ~]#
[root@CentOS-Python ~]# cd /home/python
[root@CentOS-Python python]# ls
sw1-confg  sw2-confg  sw3-confg  sw4-confg  sw5-confg
```

（4）用 cat 開啟其中任意一個 config 檔案，驗證其內容，如下圖所示。

```
[root@CentOS-Python python]# cat sw1-confg
!
! Last configuration change at 10:56:55 UTC Fri May 8 2020 by python
!
version 15.2
service timestamps debug datetime msec
service timestamps log datetime msec
no service password-encryption
service compress-config
!
hostname SW1
!
boot-start-marker
boot-end-marker
!
!
!
username python privilege 15 password 0 123
no aaa new-model
!
!
!
!
!
!
!
no ip domain-lookup
ip domain-name python.com
ip cef
no ipv6 cef
!
!
file prompt quiet
!
spanning-tree mode pvst
spanning-tree extend system-id
!
vlan internal allocation policy ascending
!
!
!
!
!
!
!
```

Python 網路運行維護實戰
（實機）

本章將以筆者在工作中遇到的 3 個實際案例示範 Python 在工作實戰中的應用，本章列出的 Python指令稿和程式將在生產網路裡的實機網路裝置上實戰執行。

❑ 實機執行環境

主機作業系統：Windows 10 上執行 CentOS 8（VMware 虛擬機器）

網路裝置：思科 2960、2960S、2960X、3750、3850 許多

網路裝置 OS 版本：思科 IOS、IOS-XE

Python 版本：3.8.2

實驗網路拓撲：與模擬器執行環境類似，不同點是執行 Python 的主機和網路裝置處在不同網段中，並且網路裝置的管理 IP 位址並不連續。

5.1 實驗 1：大規模批次修改交換機 QoS 的設定

在第 4 章的實驗 2 中提到了，要使用 Python 來批次連接管理 IP 位址不連續的網路裝置，可以把裝置的管理 IP 位址預先寫入一個文字檔，然後在程式中使用 for 循環配合 open() 函數和 readlines() 函數逐行讀取該文字檔裡的管理 IP 位址，達到循環批次登入多台網路裝置的目的。

在成功登入交換機後，我們可以配合 command.send() 來對網路裝置「發號施令」，但在前面的實例中我們都是將要輸入的指令預先寫在指令稿裡，如 command.send("conf t\n")、command.send("router eigrp 1\n") 和 command.send ("end\n") 等。這種將設定指令預先寫在指令稿裡的方法便於初學者了解和學習，在只有幾台裝置的實驗環境中常用。但是在有成千上萬台網路裝置需要管理的生產環境中，這種方法顯得很笨拙，缺乏靈活性。舉例來說，假設生產環境中有不同型號、不同作業系統、不同指令格式的裝置各 1 000 台，例如思科的 3750 和 3850 交換機，前者執行的是 IOS，後者執行的是 IOS-XE。

最近你接到任務，需要分別給這兩種交換機批次修改 QoS 的設定，因為兩者的指令格式差異極大（一個是 MLS QoS，一個是 MQC QoS），必須反覆修改 command.send() 部分的程式。如果只是簡單數行指令還好辦，一旦遇到大規模的設定，那麼這種方法的效率會很低。

解決這個問題的想法是分別建立兩個文字檔，一個用來儲存設定 3750 交換機要用的指令集，另一個用來儲存設定 3850 交換機要用到的指令集，然後在 Python 指令稿裡同樣透過 for 循環加 open() 函數來讀取兩個檔案裡的內容，達到分別給所有 3750 和 3850 交換機做 QoS 設定的目的，這樣做的好處是無須修改 command.send() 部分的程式，因為所有的命令列已經在文字檔裡預先設定好了。

但是新的問題又來了，每次配備不同型號的裝置，都必須手動修改 open() 函數所開啟的設定文字檔及 IP 位址檔案。如給 3750 交換機做設定時，需要 open('command_ 3750.txt') 和 open('ip_3750.txt')；給 3850 交換機做設定時，又需要 open('command_3850.txt') 和 open('ip_3850.txt')，這樣一來二去修改設定指令稿的做法大幅缺乏靈活性。如果只有兩種不同型號、不同指令格式的裝置還能應付，那麼當生產環境中同時使用 3750（IOS）、3850（IOS-XE）、Nexus 3k/5k/7k/9k（NX-OS）、CRS3/ASR9K（IOS-XR），甚至其他廠商的裝置，而又要對所有這些裝置同時修改某個共有的設定。例如網路新增某台 TACACS 伺服器，要統一給所有裝置修改它們的 AAA 設定；又或網路新增某台 NMS 系統，要統一給所有裝置修改 SNMP 設定。因為不同 OS 的裝置的設定指令完全不同，這時就能體會到痛苦了。此時我們可以用下面實驗中的 sys.argv 來解決這個問題。

5.1.1 實驗背景

本實驗將在實機上完成。

- 假設現在手邊有 3 台管理 IP 位址在 192.168.100.x /24 網段的 3750 交換機和 3 台管理 IP 位址在 172.16.100.x/24 網段的 3850 交換機，它們的 hostname 和管理 IP 位址分別如下。

3750_1: 192.168.100.11
3750_2: 192.168.100.22
3750_3: 192.168.100.33

3850_1: 172.16.100.11
3850_2: 172.16.100.22
3850_3: 172.16.100.33

5.1.2 實驗目的

- 修改所有 3750 和 3850 交換機的 QoS 設定，更改它們出佇列（output queue）的佇列參數集 2（queue-set 2）的快取（buffers）設定，給佇列 1、2、3 和 4 分別分配 15%、25%、40% 和 20% 的快取（預設狀況下是 25%、25%、25% 和 25%）。

5.1.3 實驗準備

（1）首先建立名為 command_3750.txt 和 ip_3750.txt 的兩個文字檔，分別用來儲存我們將要設定 3750 交換機的 QoS 指令，以及所有 3750 交換機的管理 IP 位址。

```
[root@CentOS-Python ~]# cat command_3750.txt
configure terminal
mls qos queue-set output 1 buffers 15 25 40 20
end
wr mem
[root@CentOS-Python ~]# cat ip_3750.txt
192.168.100.11
192.168.100.22
192.168.100.33
```

（2）同理，建立名為 command_3850.txt 和 ip_3850.txt 的兩個文字檔，分別用來儲存我們將要設定 3850 交換機的 QoS 指令，以及所有 3850 交換機的管理 IP 位址。

```
[root@CentOS-Python ~]# cat command_3850.txt
configure terminal
class-map match-any cos7
match cos 7
class-map match-any cos1
match cos 1
```

```
exit
policy-map queue-buffer
class cos7
bandwidth percent 10
queue-buffers ratio 15
class cos1
bandwidth percent 30
queue-buffers ratio 25
exit
exit
interface gi1/0/1
service-policy output queue-buffer
end
wr mem

[root@CentOS-Python ~]# cat ip_3850.txt
172.16.100.11
172.16.100.22
172.16.100.33
```

（3）在主機上建立實驗 1 的指令稿，將其命名為 lab1.py。

```
[root@CentOS-Python ~]#
[root@CentOS-Python ~]# vi lab1.py_
```

5.1.4 實驗程式

將下列程式寫入指令稿 lab1.py。

```
import paramiko
import time
import getpass
import sys


username = input('username: ')
```

```
password = getpass.getpass('password: ')
ip_file = sys.argv[1]
cmd_file = sys.argv[2]


iplist = open(ip_file，'r')
for line in iplist.readlines():
    ip = line.strip()
    ssh_client = paramiko.SSHClient()
    ssh_client.set_missing_host_key_policy(paramiko.AutoAddPolicy())
    ssh_client.connect(hostname=ip，username=username，password=password)
    print "You have successfully connect to "，ip
    command = ssh_client.invoke_shell()
    cmdlist = open(cmd_file，'r')
    cmdlist.seek(0)
    for line in cmdlist.readlines():
        command.send(line + "\n")
        time.sleep(1)
    cmdlist.close()
    output = command.recv(65535)
    print (output.decode("ascii"))

iplist.close()
ssh_client.close
```

5.1.5 程式分段說明

（1）因為要用到 sys.argv，所以我們匯入了 sys 模組。sys 模組是 Python 中十分常用的內建模組。其餘部分的程式都是在第 4 章中講過的，不再贅述。

```
import paramiko
import time
import getpass
import sys
```

```
username = input('username: ')
password = getpass.getpass('password: ')
```

（2）建立兩個變數 ip_file 和 cmd_file，分別對應 sys.argv[1] 和 sys.argv[2]。

```
ip_file = sys.argv[1]
cmd_file = sys.argv[2]
```

argv 是 argument variable 參數變數的簡寫形式，這個變數的傳回值是一個串列，該串列中的元素即我們在主機命令列裡執行 Python 指令稿時輸入的指令。sys.argv[0] 一般是被呼叫的 .py 指令稿的檔案名稱，從 sys.argv[1] 開始就是為這個指令稿增加的參數。舉個實例，我們現在傳回主機，輸入下面這行指令。

```
[root@CentOS-Python ~]# python3.8 lab1.py ip_3750.txt cmd_3750.txt
```

那麼，這時的 sys.argv 即含有 lab1.py、ip_3750.txt、cmd_3750.txt 3 個元素的串列。這時，sys.argv[0] = lab1.py，sys.argv[1] = ip_3750.txt，sys.argv[2] = cmd_3750.txt。對應地，程式裡的 ip_file = sys.argv[1] 此時等於 ip_file = ip_3750.txt，cmd_file = sys.argv[2] 此時等於 cmd_file = cmd_3750.txt。同理，如果這時我們在主機上執行以下指令。

```
[root@CentOS-Python ~]# python3.8 lab1.py ip_3850.txt cmd_3850.txt
```

則此時 ip_file = ip_3850.txt，cmd_file = cmd_3850.txt。由此可見，配合 sys.argv，我們可以很靈活地選用指令稿需要呼叫的參數（文字檔），而無須反反覆覆地修改指令稿程式。

（3）需要注意的是，在剩下的程式中，我們沒有在指令稿裡預先寫好實際的 QoS 設定指令，取而代之的是透過 cmd_file = sys.argv[2] 配合 cmdlist = open(cmd_file , 'r') 和 for line in cmdlist.readlines() 來讀取獨立於指令稿之外的設定指令檔案，可以隨意在命令列裡選擇我們想要的設定指令檔案，也就是本實驗中的 cmd_3750.txt 和 cmd_3850.txt。

```
iplist = open(ip_file，'r')
for line in iplist.readlines():
    ip = line.strip()
    ssh_client = paramiko.SSHClient()
    ssh_client.set_missing_host_key_policy(paramiko.AutoAddPolicy())
    ssh_client.connect(hostname=ip，username=username，password=password)
    print "You have successfully connect to "，ip
    command = ssh_client.invoke_shell()
    cmdlist = open(cmd_file，'r')
    cmdlist.seek(0)
    for line in cmdlist.readlines():
        command.send(line + "\n")
        time.sleep(1)
    cmdlist.close()
    output = command.recv(65535)
    print (output.decode("ascii"))

iplist.close()
ssh_client.close
```

5.1.6 驗證

```
[root@CentOS-Python ~]# python3.8 lab1.py ip_3750.txt cmd_3750.txt
Username: python
password:
You have successfully connect to 192.168.100.11
3750_1#conf t
3750_1(config)#mls qos queue-set output 1 buffers 15 25 40 20
3750_1(config)#end
3750_1#wr mem
Building configuration...
[OK]

You have successfully connect to 192.168.100.22
3750_2#conf t
```

```
3750_2(config)#mls qos queue-set output 1 buffers 15 25 40 20
3750_2(config)#end
3750_2#wr mem
Building configuration...
[OK]

You have successfully connect to 192.168.100.33
3750_3#conf t
3750_3(config)#mls qos queue-set output 1 buffers 15 25 40 20
3750_3(config)#end
3750_3#wr mem
Building configuration...
[OK]

[root@CentOS-Python ~]# python lab1.py ip_3850.txt cmd_3850.txt
Username: python
password:

You have successfully connect to 172.16.100.11
3850_1#configure terminal
Enter configuration commands, one per line.  End with CNTL/Z.
3850_1(config)#class-map match-any cos7
3850_1(config-cmap)#match cos 7
3850_1(config-cmap)#class-map match-any cos1
3850_1(config-cmap)#match cos 1
3850_1(config-cmap)#exit
3850_1(config)#policy-map queue-buffer
3850_1(config-pmap)#class cos7
3850_1(config-pmap-c)#bandwidth percent 10
3850_1(config-pmap-c)#queue-buffers ratio 15
3850_1(config-pmap-c)#class cos1
3850_1(config-pmap-c)#bandwidth percent 30
3850_1(config-pmap-c)#queue-buffers ratio 25
3850_1(config-pmap-c)#exit
3850_1(config-pmap)#exit
```

```
3850_1(config)#interface gi1/0/1
3850_1(config-if)#service-policy output queue-buffer
3850_1(config-if)#end
3850_1#wr mem
Building configuration...
Compressed configuration from 62654 bytes to 19670 bytes[OK]

You have successfully connect to 172.16.100.22
3850_2#configure terminal
Enter configuration commands, one per line.  End with CNTL/Z.
3850_2(config)#class-map match-any cos7
3850_2(config-cmap)#match cos 7
3850_2(config-cmap)#class-map match-any cos1
3850_2(config-cmap)#match cos 1
3850_2(config-cmap)#exit
3850_2(config)#policy-map queue-buffer
3850_2(config-pmap)#class cos7
3850_2(config-pmap-c)#bandwidth percent 10
3850_2(config-pmap-c)#queue-buffers ratio 15
3850_2(config-pmap-c)#class cos1
3850_2(config-pmap-c)#bandwidth percent 30
3850_2(config-pmap-c)#queue-buffers ratio 25
3850_2(config-pmap-c)#exit
3850_2(config-pmap)#exit
3850_2(config)#interface gi1/0/1
3850_2(config-if)#service-policy output queue-buffer
3850_2(config-if)#end
3850_2#wr mem
Building configuration...
Compressed configuration from 62654 bytes to 19670 bytes[OK]

You have successfully connect to 172.16.100.33
3850_3#configure terminal
Enter configuration commands, one per line.  End with CNTL/Z.
3850_3(config)#class-map match-any cos7
```

```
3850_3(config-cmap)#match cos 7
3850_3(config-cmap)#class-map match-any cos1
3850_3(config-cmap)#match cos 1
3850_3(config-cmap)#exit
3850_3(config)#policy-map queue-buffer
3850_3(config-pmap)#class cos7
3850_3(config-pmap-c)#bandwidth percent 10
3850_3(config-pmap-c)#queue-buffers ratio 15
3850_3(config-pmap-c)#class cos1
3850_3(config-pmap-c)#bandwidth percent 30
3850_3(config-pmap-c)#queue-buffers ratio 25
3850_3(config-pmap-c)#exit
3850_3(config-pmap)#exit
3850_3(config)#interface gi1/0/1
3850_3(config-if)#service-policy output queue-buffer
3850_3(config-if)#end
3850_3#wr mem
Building configuration...
Compressed configuration from 62654 bytes to 19670 bytes[OK]
```

5.2 實驗 2：pythonping 的使用方法

在第 3 章中，我們曾經提到過在 Python 中用來執行 ping 指令的模組有很多種，os、subprocess 及 pyping 都可以用來 ping 指定的 IP 位址或 URL。三者的區別是 os 和 subprocess 在執行 ping 指令時指令稿會將系統執行 ping 時的回應內容顯示出來，有時這些回應內容並不是必要的，例如下面是用 subprocess 模組在 CentOS 主機上執行 ping -c 3 www.cisco.com 指令的指令稿及執行指令稿後的回應內容。

```
[root@CentOS-Python ~]# cat ping.py

import subprocess
```

```
target = 'www.cisco.com'
ping_result = subprocess.call(['ping','-c','3',target])
if ping_result == 0:
    print (target + ' is reachable.')
else:
    print (target + ' is not reachable.')

[root@CentOS-Python ~]# python3.8 ping.py
PING www.cisco.com (27.151.12.183) 56(84) bytes of data.
64 bytes from 27.151.12.183 (27.151.12.183): icmp_seq=1 ttl=55 time=19.8 ms
64 bytes from 27.151.12.183 (27.151.12.183): icmp_seq=2 ttl=55 time=19.9 ms
64 bytes from 27.151.12.183 (27.151.12.183): icmp_seq=3 ttl=55 time=19.9 ms

--- www.cisco.com ping statistics ---
3 packets transmitted, 3 received, 0% packet loss, time 5ms
rtt min/avg/max/mdev = 19.765/19.875/19.935/0.077 ms
www.cisco.com is reachable.
[root@CentOS-Python ~]#
```

由上面程式可以明顯看到，上述回應內容過多，在用指令稿一次性 ping 成千上萬個 IP 位址或 URL 時非常影響美觀和閱讀，因為我們真正關心的其實是最後一句用 Python 列印出來的通知使用者目標 IP 位址或 URL 是否可達的內容 "www.cisco.com is reachable"，而用 pyping 來執行 ping 指令則不會有回應內容過多的問題。很遺憾的是，pyping 只支援 Python 2，並且截至 2020 年 5 月，作者似乎也沒有繼續更新 pyping 來支援 Python 3 的意願（雖然 pyping 依然能透過 pip3 下載 Python 3 的版本，但是使用時會顯示出錯 "ModuleNotFoundError: No module named 'core'"）。

在 Python 3.8 裡，我們可以使用 pythonping 作為 pyping 的替代品。本節實驗將詳細介紹 pythonping 的使用方法。

5.2.1 實驗背景

本實驗將在實機上完成。

- 某公司有 48 埠的思科 3750 交換機共 1 000 台，分別分佈在 5 個子網路遮罩為 /24 的 B 類別網路子網下：172.16.0.x /24，172.16.1.x /24，172.16.2.x /24，172.16.3.x /24，172.16.4.x /24。

5.2.2 實驗目的

- 在不借助任何協力廠商 NMS 軟體或網路安全工具幫助的前提下，使用 Python 指令稿依次 ping 所有交換機的管理 IP 位址，來確定目前（需要記錄下執行指令稿時的時間，要求精確到年月日和分時秒）有哪些交換機可達，並且統計目前每個交換機有多少使用者端的物理通訊埠是 Up 的（串聯通訊埠不算），以及這 1 000 台交換機所有 Up 的使用者端物理通訊埠的總數，並統計網路裡的通訊埠使用率（也就是物理通訊埠的 Up 率）。

5.2.3 實驗想法

- 根據實驗目的，我們可以寫兩個指令稿，指令稿 1 透過匯入 pythonping 模組來掃描這 5 個網段下所有交換機的管理 IP 位址，看哪些管理 IP 位址是可達的。因為子網路遮罩是 255.255.255.0，表示每個網段下的管理 IP 位址的前三位都是固定不變的，只有最後一位會在 1 ～ 254 中變化。我們可以在第一個指令稿中使用 for 循環來 ping .1 到 .254，然後將所有該網段下可達的交換機管理 IP 位址都寫入並儲存在一個名為 reachable_ip.txt 的文字檔中。

- 因為這裡有 5 個連續的 /24 的網段需要掃描（從 172.16.0.x 到 172.16.4.x），我們可以在指令稿 1 中再寫一個 for 循環來連續 ping 這 5 個網段，然後把上一個 ping .1 到 .254 的 for 循環嵌入這一個 for 循環，這樣就能讓 Python 一次性把 5 個 /24 網段下總共 1270 個可用 IP 位址（254×5 = 1270）全部 ping 一遍。

■ 在用指令稿 1 產生 reachable_ip.txt 檔案後，我們可以再寫一個指令稿 2 來讀取該文字檔中所有可達的交換機的管理 IP 位址，依次登入所有這些可達的交換機，輸入指令 show ip int brief | i up 檢視有哪些通訊埠是 Up 的，再配合正規表示法（re 模組），在回應內容中比對我們所要的使用者端物理通訊埠編號（Gix/x/x），統計它們的總數，即可獲得目前一個交換機有多少個物理通訊埠是 Up 的。（註：因為 show ip int brief | i up 的回應內容裡也會出現 10GB 的串聯通訊埠 Tex/x/x 及虛擬通訊埠，例如 VLAN 或 loopback 通訊埠，所以這裡強調的是用正規表示法來比對使用者端物理通訊埠 Gix/x/x）。

5.2.4 實驗準備 - 指令稿 1

（1）pyping 為協力廠商模組，使用前需要透過 pip 下載安裝，安裝完成後進入 Python 3.8 編輯器，如果 import pythonping 沒有顯示出錯，則說明安裝成功，如下圖所示。

```
[root@CentOS-Python lab5]# pip3.8 install pythonping
Collecting pythonping
  Using cached https://files.pythonhosted.org/packages/6a/5a/f1eb3f521dcf1d2927b9be15e58a7
Installing collected packages: pythonping
  Running setup.py install for pythonping ... done
Successfully installed pythonping-1.0.9
WARNING: You are using pip version 19.2.3, however version 20.1 is available.
You should consider upgrading via the 'pip install --upgrade pip' command.
[root@CentOS-Python lab5]# python3.8
Python 3.8.2 (default, Apr 27 2020, 23:06:10)
[GCC 8.3.1 20190507 (Red Hat 8.3.1-4)] on linux
Type "help", "copyright", "credits" or "license" for more information.
>>> import pythonping
>>>
```

（2）在主機上建立一個新的資料夾，取名為 lab2，在該資料夾下建立實驗 2 的指令稿 1 檔案 lab2_1.py，如下圖所示。

```
[root@CentOS-Python ~]# mkdir lab2
[root@CentOS-Python ~]# cd lab2
[root@CentOS-Python lab2]# vi lab2_1.py
```

5.2.5 實驗程式 - 指令稿 1

將下列程式寫入指令稿 lab2_1.py。

```python
from pythonping import ping
import os

if os.path.exists('reachable_ip.txt'):
    os.remove('reachable_ip.txt')

third_octet = range(5)
last_octet = range(1, 255)

for ip3 in third_octet:
    for ip4 in last_octet:
        ip = '172.16.' + str(ip3) + '.' + str(ip4)
        ping_result = ping(ip)
        f = open('reachable_ip.txt', 'a')
        if 'Reply' in str(ping_result):
            print (ip + ' is reachable. ')
            f.write(ip + "\n")
        else:
            print (ip + ' is not reachable. ')

f.close()
```

5.2.6 指令稿 1 程式分段說明

（1）我們匯入 pythonping 和 os 兩個模組。在 pythonping 中，最核心的函數顯然是 ping()，因為 pythonping 的模組名稱長度偏長，這裡用 from…import…將 ping() 函數匯入後，後面呼叫時就能省去使用 pythonping.ping()

完整函數路徑的麻煩，體會到直接使用 ping() 的便利。至於 os 模組的用法下面會講到。

```
from pythonping import ping
import os
```

（2）每次我們執行指令稿 1，都不希望保留上一次執行指令稿時產生的 reachable_ip.txt 檔案，因為在有成千上萬台裝置的大型網路裡，每時每刻可達交換機的數量都有可能改變。這時可以用 os 模組下的 os.path.exists() 方法來判斷該檔案是否存在，如果存在，則用 os.remove() 方法將該檔案刪除。這樣可以確保每次執行指令稿 1 時，reachable_ip.txt 檔案中只會包含本次執行指令稿後所有可達的交換機管理 IP 位址。

```
if os.path.exists('reachable_ip.txt'):
    os.remove('reachable_ip.txt')
```

（3）實驗想法中講到，5 個 /24 網段的管理 IP 位址是有規律可循的，頭兩位都不變，為 172.16，第三位為 0 ～ 4，第四位為 1 ～ 254，因此可以透過 range(5) 和 range(1，255) 分別建立兩個整數串列來囊括管理 IP 位址的第三位和第四位，為後面的兩個 for 循環做準備。

```
third_octet = range(5)
last_octet = range(1, 255)
```

（4）透過兩個 for 循環做巢狀結構，依次從 172.16.0.1、172.16.0.2、172.16.0.3……一直檢查到 172.16.4.254 為止，然後配合 ping() 函數來依次 ping 所有這些管理 IP 位址（前面已經提到 ping() 函數是透過 from pythonping import ping 從 pythonping 模組匯入的）。

```
for ip3 in third_octet:
    for ip4 in last_octet:
        ip = '172.16.' + str(ip3) + '.' + str(ip4)
        ping_result = ping(ip)
```

（5）在實驗 2 的指令稿裡，我們將所有可達的管理 IP 位址以追加模式（a）寫入 reachable_ip.txt 檔案。因為 pythonping 模組在執行過程中不顯示任何回應內容，就無法知道指令稿執行的進度，所以透過 print (ip + ' is reachable. ') 和 print (ip + ' is not reachable. ') 分別列印出目標 IP 位址是否可達的資訊。

```
f = open('reachable_ip.txt','a')
if 'Reply' in str(ping_result):
    print (ip + ' is reachable. ')
    f.write(ip + "\n")
else:
    print (ip + ' is not reachable. ')
```

這裡重點解釋 if 'Reply' in str(ping_result): 的用法和原理。

在使用 os、subprocess 和 pyping 等模組做 ping 測試時，如果目標 IP 位址可達，則它們會傳回整數 0；如果無法連接，則傳回非 0 的整數。而 pythonping 不同，在 pythonping 中，ping() 函數預設對目標 IP 位址 ping 4 次，當目標 IP 位址可達時，ping() 函數傳回的是 "Reply from x.x.x.x, x bytes in xx.xx ms"；如果無法連接，則傳回的是 "Request timed out"，如下圖所示。

```
[root@CentOS-Python lab5]# python3.8
Python 3.8.2 (default, Apr 27 2020, 23:06:10)
[GCC 8.3.1 20190507 (Red Hat 8.3.1-4)] on linux
Type "help", "copyright", "credits" or "license" for more information.
>>> from pythonping import ping
>>> ping('www.cisco.com')
Reply from 117.34.19.157, 9 bytes in 36.67ms
Reply from 117.34.19.157, 9 bytes in 36.09ms
Reply from 117.34.19.157, 9 bytes in 36.71ms
Reply from 117.34.19.157, 9 bytes in 41.84ms

Round Trip Times min/avg/max is 36.09/37.83/41.84 ms
>>> ping('10.1.1.1')
Request timed out
Request timed out
Request timed out
Request timed out

Round Trip Times min/avg/max is 2000/2000.0/2000 ms
>>>
```

也許你會問：既然 pythonping 的 ping() 函數傳回的不再是 0 或非 0 的整
數，那麼我們怎麼將上面代表目標 IP 位址可達的 "Reply from x.x.x.x, x
bytes in xx.xx ms" 和代表目標 IP 位址無法連接的 "Request timed out" 透過
if 敘述將可達的目標 IP 位址列印出來並寫入 reachable_ip.txt 檔案呢？也
許你猜到用成員運算子 in 來判斷傳回值中是否有 "Reply" 和 "Request" 這
兩個字串。如果有 "Reply"，則說明目標可達；如果沒有，則說明目標無
法連接，但前提是 ping() 函數傳回值的類型必須是字串。而實際情況是
ping() 函數傳回值的類型是一個叫作 pythonping.executor.ResponseList 的特
殊類型，如下圖所示。

```
>>>
>>> from pythonping import ping
>>> ping_result = ping('www.cisco.com')
>>> type(ping_result)
<class 'pythonping.executor.ResponseList'>
>>>
```

因此，必須透過 str() 函數將它轉換成字串後才能使用成員運算子 in 來作
判斷（即這裡的 str(ping_result)，否則 if 'Reply' in str(ping_result): 會永遠
傳回布林值 False，表示目標無法連接。

（6）最後在退出程式前不要忘記關閉已經開啟的 reachable_ip.txt 檔案。

```
f.close()
```

5.2.7 指令稿 1 驗證

（1）執行指令稿 1 前，確認 /root/lab2 資料夾下只有 lab2_1.py 這一個檔
案。

```
[root@CentOS-Python lab2]# ls
lab2_1.py
[root@CentOS-Python lab2]#
```

（2）執行指令稿 1，因為 1270 個管理 IP 位址實在太多，不方便畫面示範，所以這裡將 last_octet = range(1，255) 改為 last_octet = range(1，4)，只 ping 每個網段下前 3 個管理 IP 位址，也就是總共 15 個管理 IP 位址。執行指令稿後的回應內容如下圖所示。

```
[root@CentOS-Python lab2]# python3.8 lab2_1.py
172.16.0.1 is reachable.
172.16.0.2 is not reachable.
172.16.0.3 is not reachable.
172.16.1.1 is not reachable.
172.16.1.2 is not reachable.
172.16.1.3 is not reachable.
172.16.2.1 is not reachable.
172.16.2.2 is not reachable.
172.16.2.3 is not reachable.
172.16.3.1 is not reachable.
172.16.3.2 is not reachable.
172.16.3.3 is not reachable.
```

（3）再次檢視 /root/lab2 資料夾，可以看到這時多出來一個 reachable_ip.txt 的文字檔，該檔案正是指令稿自動產生的，用來儲存所有可達交換機的管理 IP 位址。可以看到只有 172.16.0.1 這一個管理 IP 位址被寫入 reachable_ip.txt，至此證明指令稿 1 執行成功，如下圖所示。

```
[root@CentOS-Python lab2]# ls
lab2_1.py  reachable_ip.txt
[root@CentOS-Python lab2]# cat reachable_ip.txt
172.16.0.1
[root@CentOS-Python lab2]# _
```

5.2.8 實驗準備 - 指令稿 2

（1）説明指令稿 2 之前，先來看下在一個 48 埠的思科 3750 交換機裡輸入指令 show ip int brief | i up 後能獲得什麼樣的回應內容，如下圖所示。

```
                    #show ip int b | i up
Vlan3999                              YES NVRAM  up                      up
GigabitEthernet1/0/1   unassigned     YES unset  up                      up
GigabitEthernet1/0/3   unassigned     YES unset  up                      up
GigabitEthernet1/0/5   unassigned     YES unset  up                      up
GigabitEthernet1/0/6   unassigned     YES unset  up                      up
GigabitEthernet1/0/7   unassigned     YES unset  up                      up
GigabitEthernet1/0/9   unassigned     YES unset  up                      up
GigabitEthernet1/0/10  unassigned     YES unset  up                      up
GigabitEthernet1/0/11  unassigned     YES unset  up                      up
GigabitEthernet1/0/12  unassigned     YES unset  up                      up
GigabitEthernet1/0/15  unassigned     YES unset  up                      up
GigabitEthernet1/0/16  unassigned     YES unset  up                      up
GigabitEthernet1/0/17  unassigned     YES unset  up                      up
GigabitEthernet1/0/18  unassigned     YES unset  up                      up
GigabitEthernet1/0/20  unassigned     YES unset  up                      up
GigabitEthernet1/0/21  unassigned     YES unset  up                      up
GigabitEthernet1/0/22  unassigned     YES unset  up                      up
GigabitEthernet1/0/29  unassigned     YES unset  up                      up
GigabitEthernet1/0/31  unassigned     YES unset  up                      up
GigabitEthernet1/0/33  unassigned     YES unset  up                      up
GigabitEthernet1/0/34  unassigned     YES unset  up                      up
GigabitEthernet1/0/36  unassigned     YES unset  up                      up
GigabitEthernet1/0/37  unassigned     YES unset  up                      up
GigabitEthernet1/0/38  unassigned     YES unset  up                      up
GigabitEthernet1/0/39  unassigned     YES unset  up                      up
GigabitEthernet1/0/42  unassigned     YES unset  up                      up
GigabitEthernet1/0/45  unassigned     YES unset  up                      up
Te1/0/1                unassigned     YES unset  up                      up
Te1/0/2                unassigned     YES unset  up                      up
```

由上面可以看到，除了 GigabitEthernet 使用者端物理通訊埠，還有 VLAN
虛擬通訊埠和兩個 10GB 的串聯通訊埠 Te1/0/1 和 Te1/0/2。在實驗目的中
已經明確説明不考慮虛擬通訊埠和串聯通訊埠，只統計共有多少個使用者
端物理通訊埠是 Up 的。

（2）建立實驗 2 的第二個指令稿，取名為 lab2_2.py，如下圖所示。

```
[root@CentOS-Python lab2]#
[root@CentOS-Python lab2]# vi lab2_2.py
```

5.2.9 實驗程式 - 指令稿 2

將下列程式寫入指令稿 lab2_2.py。

```
import paramiko
import time
import re
```

```
from datetime import datetime
import socket
import getpass

username = input('Enter your SSH username: ')
password = getpass.getpass('Enter your SSH password: ')
now = datetime.now()
date = "%s-%s-%s" % (now.month，now.day，now.year)
time_now = "%s:%s:%s" % (now.hour，now.minute，now.second)

switch_with_tacacs_issue = []
switch_not_reachable = []
total_number_of_up_port = 0

iplist = open('reachable_ip.txt')
number_of_switch = len(iplist.readlines())
total_number_of_ports = number_of_switch * 48

iplist.seek(0)
for line in iplist.readlines():
    try:
        ip = line.strip()
        ssh_client = paramiko.SSHClient()
        ssh_client.set_missing_host_key_policy(paramiko.AutoAddPolicy())
        ssh_client.connect(hostname=ip，username=username，password=password)
        print ("\nYou have successfully connect to "，ip)
        command = ssh_client.invoke_shell()
        command.send('term len 0\n')
        command.send('show ip int b | i up\n')
        time.sleep(1)
        output = command.recv(65535)
        #print (output)
        search_up_port = re.findall(r'GigabitEthernet'，output)
        number_of_up_port = len(search_up_port)
        print (ip + " has " + str(number_of_up_port) + " ports up.")
        total_number_of_up_port += number_of_up_port
```

```
    except Paramiko.ssh_exception.AuthenticationException:
        print ("TACACS is not working for " + ip + ".")
        switch_with_tacacs_issue.append(ip)
    except socket.error:
        print (ip + " is not reachable.")
        switch_not_reachable.append(ip)
iplist.close()

print ("\n")
print ("There are totally " + str(total_number_of_ports) + " ports available
in the network.")
print (str(total_number_of_up_port) + " ports are currently up.")
print ("Port up rate is %.2f%%" % (total_number_of_up_port / float(total_
number_of_ports) * 100))
print ('\nTACACS is not working for below switches: ')
for i in switch_with_tacacs_issue:
    print (i)
print ('\nBelow switches are not reachable: ')
for i in switch_not_reachable:
    print (i)
f = open(date + ".txt"，"a+")
f.write('As of ' + date + " " + time_now)
f.write("\n\nThere are totally " + str(total_number_of_ports) + " ports
available in the network.")
f.write("\n" + str(total_number_of_up_port) + " ports are currently up.")
f.write("\nPort up rate is %.2f%%" % (total_number_of_up_port / float(total_
number_of_ports) * 100))
f.write("\n************************************************************\n\n")
f.close()
```

5.2.10 指令稿 2 程式分段說明

（1）這裡匯入了 datetime 模組，這個 Python 內建模組可以用來顯示執行指令稿時的系統日期和時間，因為實驗目的裡提到需要記錄下執行指令稿時的時間（精確到年月日和分時秒）。其餘模組的用法在前面已經都講過，這裡不再贅述。

```
import paramiko
import time
import re
from datetime import datetime
import socket
import getpass
```

（2）使用 input() 和 getpass.getpass() 來分別提示使用者輸入 SSH 登入交換機的使用者名稱和密碼。記錄目前時間可以呼叫 datetime.now() 方法，我們將它設定值給變數 now。datetime.now() 方法下面又包含了 .year()（年）、.month()（月）、.day()（日）、.hour()（時）、.minute()（分）、.second()（秒）幾個子方法，這裡將「月 - 日 - 年」設定值給變數 date，將「時 : 分 : 秒」設定值給變數 time_now。

```
username = input('Enter your SSH username: ')
password = getpass.getpass('Enter your SSH password: ')
now = datetime.now()
date = "%s-%s-%s" % (now.month，now.day，now.year)
time_now = "%s:%s:%s" % (now.hour，now.minute，now.second)
```

（3）建立 switch_with_tacacs_issue 和 switch_not_reachable 兩個空串列來統計有哪些交換機的 TACACS 故障導致使用者驗證失敗，哪些交換機的管理 IP 位址無法連接。另外建立一個 total_number_of_up_port 變數，將其初值設為 0，再用累加的方法統計所有可達的 3750 交換機上狀態為 Up 的通訊埠的總數。

```
switch_with_tacacs_issue = []
switch_not_reachable = []
total_number_of_up_port = 0
```

（4）用 open() 函數開啟指令稿 1 建立的 reachable_ip.txt 檔案，用 readlines()
將其內容以串列形式傳回，再配合 len() 函數獲得可達交換機的數量，因為
每個交換機都有 48 個通訊埠，所以透過交換機數量 ×48 可以獲得通訊埠
總數（無論通訊埠狀態是否為 Up）。

```
iplist = open('reachable_ip.txt')
number_of_switch = len(iplist.readlines())
total_number_of_ports = number_of_switch * 48
```

（5）因為已經用 open() 函數開啟過一次 reachable_ip.txt 檔案，所以要用
seek(0) 回到檔案的起始位置。這裡需要注意用來應對交換機登入失敗的問
題，異常處理敘述 try 要寫在 for 循環的下面，剩下的程式是使用 Paramiko
配合 for 循環登入交換機並進入命令列的最基礎的基礎知識，這裡不再贅
述。

```
iplist.seek(0)
for line in iplist.readlines():
    try:
        ip = line.strip()
        ssh_client = Paramiko.SSHClient()
        ssh_client.set_missing_host_key_policy(Paramiko.AutoAddPolicy())
        ssh_client.connect(hostname=ip，username=username，password=password)
        print ("\nYou have successfully connect to "，ip)
        command = ssh_client.invoke_shell()
```

（6）因為是 48 埠的交換機，show ip int brief | i up 的回應內容會比較長，
無法一次性完整地顯示，所以首先要用指令 term len 0 完整地顯示所有的
回應內容。"sleep" 1s 後再用 recv(65535) 將所有的回應內容儲存在變數
output 中，如果你想在指令稿執行過程中檢視完整的回應內容，可以選擇
print (output)；如果不想看，則在其前面加上註釋符號 #。

```
        command.send('term len 0\n')
        command.send('show ip int b | i up\n')
        time.sleep(1)
```

```
output = command.recv(65535)
#print (output)
```

（7）因為我們只想統計有多少個使用者端的物理通訊埠（GigabitEthernet）是 Up 的，所以可以用正規表示法的 findall() 方法去精確比對 GigabitEthernet，將 findall() 傳回的串列設定值給變數 search_up_port，然後透過 len(search_up_port) 即可獲得 Up 的物理通訊埠的數量，並將該數量設定值給變數 number_of_port。隨後列印出每個交換機有多少個使用者端物理通訊埠是 Up 的。因為前面已經定義了變數 total_number_of_up_port，並將整數 0 設定值給它，所以可以透過 total_number_of_up_port += number_of_up_port 的方法將每個交換機 Up 的物理通訊埠數量累加起來，最後獲得整個網路下 Up 的物理通訊埠的總數。後面的異常處理敘述 except 和關閉檔案的部分不再贅述。

```
        search_up_port = re.findall(r'GigabitEthernet', output)
        number_of_up_port = len(search_up_port)
        print (ip + " has " + str(number_of_up_port) + " ports up.")
        total_number_of_up_port += number_of_up_port
    except Paramiko.ssh_exception.AuthenticationException:
        print ("TACACS is not working for " + ip + ".")
        switch_with_tacacs_issue.append(ip)
    except socket.error:
        print (ip +  " is not reachable.")
        switch_not_reachable.append(ip)
 iplist.close()
```

（8）最後，除了將各種統計資訊列印出來，我們還將另外建立一個檔案，透過 f = open(date + ".txt" , "a + ") 將執行指令稿時的日期用作該指令稿的名字，將統計資訊寫入，方便以後調閱檢視。注意寫入的內容裡有 f.write('As of ' + date + " " + time_now)，這樣可以清晰直觀地看到我們是在哪一天的幾時幾分幾秒執行的指令稿。為什麼要用日期作為檔案名稱呢？這樣做的好處是一旦執行指令稿時的日期不同，指令稿就會自動建立一個新的檔案，例如 2018 年 6 月 16 日執行了一次指令稿，Python 建立了一個名為 6-16-2018.txt 的檔案，如果第二天再執行一次指令稿，Python 又會建立一個名為 6-17-2018.txt 的檔案。如果在同一天裡數次執行指令稿，則多次執行的結果會以追加的形式寫入同一個 .txt 檔案，不會建立新檔案。這麼做可以配合 Windows 的 Task Scheduler 或 Linux 的 Crontab 來定期自動執行指令稿，每天自動產生當天的通訊埠使用量的統計情況，方便公司管理層隨時觀察網路裡交換機的通訊埠使用情況。

```
print ("\n")
print ("There are totally " + str(total_number_of_ports) + " ports available
in the network.")
print (str(total_number_of_up_port) + " ports are currently up.")
print ("Port up rate is %.2f%%" % (total_number_of_up_port / float(total_
number_of_ports) * 100))
print ('\nTACACS is not working for below switches: ')
for i in switch_with_tacacs_issue:
    print (i)
print ('\nBelow switches are not reachable: ')
for i in switch_not_reachable:
    print (i)
f = open(date + ".txt"，"a+")
f.write('As of ' + date + " " + time_now)
f.write("\n\nThere are totally " + str(total_number_of_ports) + " ports
available in the network.")
f.write("\n" + str(total_number_of_up_port) + " ports are currently up.")
f.write("\nPort up rate is %.2f%%" % (total_number_of_up_port / float(total_
number_of_ports) * 100))
f.write("\n*************************************************************\n\n")
f.close()
```

5.2.11 指令稿 2 驗證

（1）移動到 /root/lab2，執行指令稿 lab2_2.py，讓指令稿讀取 reachable_ip.txt 檔案自動登入所有可達的交換機（出於示範的目的，這裡將之前除

172.16.0.1 外所有無法連接的交換機全部開啟，並將它們的管理 IP 位址寫入 reachable_ip.txt），並檢視每個交換機目前有多少個通訊埠是 Up 的，最後列出統計資料，如下圖所示。

```
[root@localhost lab2]# python lab2_2.py
Enter your SSH username: parry
Enter your SSH password:

You have successfully connect to  172.16.0.1
172.16.0.1 has 27 ports up.

You have successfully connect to  172.16.0.2
172.16.0.2 has 41 ports up.

You have successfully connect to  172.16.0.3
172.16.0.3 has 41 ports up.

You have successfully connect to  172.16.1.1
172.16.1.1 has 11 ports up.

You have successfully connect to  172.16.1.2
172.16.1.2 has 15 ports up.

You have successfully connect to  172.16.1.3
172.16.1.3 has 12 ports up.

You have successfully connect to  172.16.2.1
172.16.2.1 has 7 ports up.

You have successfully connect to  172.16.2.3
172.16.2.3 has 8 ports up.

You have successfully connect to  172.16.3.1
172.16.3.1 has 48 ports up.

You have successfully connect to  172.16.3.2
172.16.3.2 has 51 ports up.

You have successfully connect to  172.16.3.3
172.16.3.3 has 44 ports up.

You have successfully connect to  172.16.4.1
172.16.4.1 has 118 ports up.

You have successfully connect to  172.16.4.2
172.16.4.2 has 65 ports up.

You have successfully connect to  172.16.4.3
172.16.4.3 has 120 ports up.


There are totally 672 ports available in the network.
608 ports are currently up.
Port up rate is 90.48%
```

（2）再次檢視 /root/lab2 資料夾，可以看到這時多出來一個名為 4-22-2019.txt 的文字檔，該檔案是指令稿 2 自動產生的，用來儲存每次執行指令稿後的統計資訊。檔案名稱即執行該指令稿的日期。在 Linux 中配合 Crontab，在 Windows Server 中配合任務計畫程式（Task Scheduler）定期每天執行該指令稿，即可達到每天自動化監控交換機通訊埠使用率的目的，如下圖所示。

```
[root@localhost lab2]# ls -l
total 16
-rw-r--r-- 1 root root  195 Apr 22 08:40 4-22-2019.txt
-rw-r--r-- 1 root root  513 Apr  6 17:38 lab2_1.py
-rw-r--r-- 1 root root 2504 Apr 15 12:56 lab2_2.py
-rw-r--r-- 1 root root  154 Apr 22 08:39 reachable_ip.txt
[root@localhost lab2]# cat 4-22-2019.txt
As of 4-22-2019 8:39:44

There are totally 672 ports available in the network.
608 ports are currently up.
Port up rate is 90.48%
*******************************************************************
[root@localhost lab2]#
```

5.3 實驗 3：利用 Python 指令稿檢查交換機的設定

前面幾個實驗（包含第 4 章）已經由淺入深地說明了 Python 在網路運行維護中的實際應用技巧，並列出了對應的範例。這裡將繼續討論如何使用 Python 來解決在大型生產網路中常見的網路運行維護自動化的需求。

5.3.1 實驗背景

本實驗將在實機上完成。

■ 某公司有 1000 台思科 2960、100 台思科 2960S、300 台思科 2960X 交換機，均為 24 埠，它們的型號和 IOS 版本分別如下表所示。

類別	型號	IOS 版本
2960	WS-C2960-24PC-L	c2960-lanbasek9-mz.122-55.SE5
2960S	WS-C2960S-F24PS-L	c2960s-universalk9-mz.150-2.SE5
2960X	WS-C2960X-24PS-L	c2960x-universalk9-mz.152-2.E5

最近公司決定將所有上述交換機的 IOS 版本升級，修補漏洞，消除 Bug。升級後的 IOS 版本分別如下。

2960：c2960-lanbasek9-mz.122-55.SE12
2960S：c2960s-universalk9-mz.150-2.SE11
2960X：c2960x-universalk9-mz.152-2.E8

5.3.2 實驗目的

■ 所有交換機的新 IOS 版本已經手動完成上傳，並且 boot system 的路徑也改成了新的 IOS 版本，為確保 IOS 版本升級順利完成，需要建立 Python 指令稿檢查所有交換機的設定是否正確，避免升級 IOS 版本的過程中出現人為的錯誤。

5.3.3 實驗想法

升級交換機 IOS 版本的驗證部分可以分為以下兩個步驟。

1. 重新啟動交換機前的驗證步驟

將新版 IOS 上傳到 flash 後到重新啟動交換機前可能會遇到以下 3 種人為錯誤。

（1）IOS 版本和交換機型號貨不對板，例如把 2960 交換機的 IOS 版本上傳給了 2960X 交換機。

（2）boot system 忘記修改或修改錯誤。

（3）重新啟動交換機前，忘記 write memory 儲存設定。

因此，在重新啟動所有交換機前，需要檢驗交換機的 3 處設定。

（1）show inventory | i PID: WS。

（2）show flash: | i c2960。

上述兩步的目的是檢視交換機型號和快閃記憶體的 IOS 版本，避免前者和後者貨不對板的情況發生。

（3）show boot | i BOOT path。

這一步的目的是檢查 boot system 設定是否修改正確。

2. 重新啟動交換機後的驗證步驟

重新啟動交換機後的驗證步驟較為簡單，只需要用 show ver | b SW Version 驗證交換機的 IOS 版本是否已經成功升級到新版本即可。

綜上所述，我們可以分別寫兩個指令稿來對應上面兩個驗證步驟。指令稿 1 用作重新啟動交換機前的驗證，指令稿 2 用作重新啟動交換機後的驗證。

5.3.4 實驗準備 - 指令稿 1

（1）在主機上建立一個新的資料夾，取名為 lab3，在該資料夾下建立實驗 3 的指令稿 1 檔案 lab3_1.py，如下圖所示。

```
[root@CentOS-Python ~]#
[root@CentOS-Python ~]# mkdir lab3
[root@CentOS-Python ~]# cd lab3
[root@CentOS-Python lab3]# vi lab3_1.py
```

（2）在主機上建立一個 ip_list.txt 檔案，該檔案用來儲存我們要登入驗證 IOS 版本是否升級成功的所有交換機的管理 IP 位址，如下圖所示。

```
[root@CentOS-Python lab3]# cat ip_list.txt
172.16.206.39
172.16.206.40
172.16.206.41
172.16.206.42
172.16.206.43
172.16.206.119
172.16.206.128
172.16.206.129
172.16.206.145
172.16.206.146
[root@CentOS-Python lab3]#
```

5.3.5 實驗程式 - 指令稿 1

將下列程式寫入指令稿 lab3_1.py。

```python
import paramiko
import time
import getpass
import sys
import re
import socket

username = input("Username: ")
password = getpass.getpass("Password: ")
iplist = open('ip_list.txt','r+')

switch_upgraded = []
switch_not_upgraded = []
switch_with_tacacs_issue = []
switch_not_reachable = []

for line in iplist.readlines():
    try:
        ip_address = line.strip()
        ssh_client = paramiko.SSHClient()
        ssh_client.set_missing_host_key_policy(paramiko.AutoAddPolicy())
        ssh_client.connect(hostname=ip_address,username=username,
```

```
password=password)
        print ("Successfully connect to "，ip_address)
        command = ssh_client.invoke_shell(width=300)
        command.send("show inventory | i PID: WS\n")
        time.sleep(0.5)
        command.send("show flash: | i c2960\n")
        time.sleep(0.5)
        command.send("show boot | i BOOT path\n")
        time.sleep(0.5)
        output = command.recv(65535)
        command.send("wr mem\n")
        switch_model = re.search(r'WS-C2960\w?-\w{4，5}-L'，output)
        ios_version = re.search(r'c2960\w?-\w{8，10}\d?-mz.\d{3}-\d{1，2}.\
w{2，4}(.bin)?'，output)
        boot_system = re.search(r'flash:.+mz.\d{3}-\d{1，2}\.\w{2，4}\.bin'，
output)
        if switch_model.group() == "WS-C2960-24PC-L" and ios_version.
group() == "c2960-lanbasek9-mz.122-55.SE12.bin" and boot_system.group() ==
'flash:c2960-lanbasek9-mz.122-55.SE12.bin' or boot_system.group() == 'flash:/
c2960-lanbasek9-mz.122-55.SE12.bin':
            switch_upgraded.append(ip_address)
        elif switch_model.group() == "WS-C2960S-F24PS-L" and ios_version.
group() == "c2960s-universalk9-mz.150-2.SE11.bin" and boot_system.group()
== 'flash:c2960s-universalk9-mz.150-2.SE11.bin' or boot_system.group() ==
'flash:/c2960s-universalk9-mz.150-2.SE11.bin':
            switch_upgraded.append(ip_address)
        elif switch_model.group() == "WS-C2960X-24PS-L" and ios_version.
group() == "c2960x-universalk9-mz.152-2.E8.bin" and boot_system.group() ==
'flash:c2960x-universalk9-mz.152-2.E8.bin' or boot_system.group() == 'flash:/
c2960x-universalk9-mz.152-2.E8.bin':
            switch_upgraded.append(ip_address)
        else:
            switch_not_upgraded.append(ip_address)
    except paramiko.ssh_exception.AuthenticationException:
        print ("TACACS is not working for " + ip_address + ".")
```

```
        switch_with_tacacs_issue.append(ip_address)
    except socket.error:
        print (ip_address +  " is not reachable.")
        switch_not_reachable.append(ip_address)

iplist.close()
ssh_client.close

print ('\nTACACS is not working for below switches: ')
for i in switch_with_tacacs_issue:
    print (i)

print ('\nBelow switches are not reachable: ')
for i in switch_not_reachable:
    print (i)

print ('\nBelow switches IOS version are up-to-date: ')
for i in switch_upgraded:
    print (i)

print ('\nBelow switches IOS version are not updated yet: ')
for i in switch_not_upgraded:
    print (i)
```

5.3.6 指令稿 1 程式分段說明

（1）在程式開始的部分匯入模組，建立 4 個空串列，用來統計有多少台交換機已經成功升級 IOS 版本，有多少台交換機沒有升級 IOS 版本，有哪些交換機因為 TACACS 或管理 IP 位址無法連接而無法登入，以及配合 for loop 和 Paramiko 依次循環 SSH 登入所有交換機，這些部分都是老生常談的話題，這裡不再贅述。

```
import paramiko
import time
```

```
import getpass
import sys
import re
import socket

username = input("Username: ")
password = getpass.getpass("Password: ")
iplist = open('ip_list.txt','r+')

switch_upgraded = []
switch_not_upgraded = []
switch_with_tacacs_issue = []
switch_not_reachable = []

for line in iplist.readlines():
    try:
        ip_address = line.strip()
        ssh_client = paramiko.SSHClient()
        ssh_client.set_missing_host_key_policy(paramiko.AutoAddPolicy())
        ssh_client.connect(hostname=ip_address,username=username,password=
password)
        print ("Successfully connect to ", ip_address)
```

（2）因為後面要用 show flash: | i c2960 檢視快閃記憶體下面的 IOS 檔案，而 IOS 檔案名稱通常很長，如果不調整寬度，則會導致後面截取的 output 不完整，進一步影響正規表示法對關鍵字的比對。所以可以用 command = ssh_client.invoke_shell(width=300) 來調整 Paramiko 回應內容的寬度（預設為 100）。

```
command = ssh_client.invoke_shell(width=300)
```

（3）command.send("show inventory | i PID: WS\n")、command.send("show flash: | i c2960\n") 和 command.send("show boot | i BOOT path\n") 這 3 個指令的作用前面已經講過，這裡給大家看看在生產網路下的 2960、2960S 和

2960X 交換機中執行這 3 行指令獲得的輸出結果（均為 IOS 上傳完畢，重新啟動交換機前的輸出結果），如下所示。

2960 交換機

```
AS-2960-EC3-G-3630#show inventory | i PID: WS
PID: WS-C2960-24PC-L  , VID: V02 ,
AS-2960-EC3-G-3630#
AS-2960-EC3-G-3630#
AS-2960-EC3-G-3630#show flash: | i c2960
    2  -rwx    9827106   Mar 4 2018 09:32:04 +03:00  c2960-lanbasek9-mz.122-55.SE12.bin
    4  drwx        512   Mar 1 1993 03:08:37 +03:00  c2960-lanbase-mz.122-44.SE2
AS-2960-EC3-G-3630#
AS-2960-EC3-G-3630#
AS-2960-EC3-G-3630#show boot | i BOOT path
BOOT path-list      : flash:c2960-lanbasek9-mz.122-55.SE12.bin
AS-2960-EC3-G-3630#
```

2960S 交換機

```
AS-2960-EC3-G-3748#show inventory | i PID: WS
PID: WS-C2960S-F24PS-L , VID: V01 ,
AS-2960-EC3-G-3748#
AS-2960-EC3-G-3748#
AS-2960-EC3-G-3748#show flash: | i c2960
    4  -rwx   14572032   May 10 2018 14:48:49 +03:00  c2960s-universalk9-mz.150-2.SE11.bin
    7  drwx        512   Mar 1 1993 03:16:11 +03:00  c2960s-universalk9-mz.150-2.SE5
AS-2960-EC3-G-3748#
AS-2960-EC3-G-3748#
AS-2960-EC3-G-3748#show boot | i BOOT path
BOOT path-list      : flash:c2960s-universalk9-mz.150-2.SE11.bin
```

2960X 交換機

```
AS-2960-EC3-G-3622#show inventory | i PID: WS
PID: WS-C2960X-24PS-L , VID: V04 ,
AS-2960-EC3-G-3622#
AS-2960-EC3-G-3622#
AS-2960-EC3-G-3622#show flash: | i c2960
    5  -rwx   21287936   Apr 9 2018 15:40:23 +03:00  c2960x-universalk9-mz.152-2.E8.bin
AS-2960-EC3-G-3622#
AS-2960-EC3-G-3622#
AS-2960-EC3-G-3622#show boot | i BOOT path
BOOT path-list      : flash:/c2960x-universalk9-mz.152-2.E8.bin
AS-2960-EC3-G-3622#
```

從上面的 show inventory | i PID: WS 的回應內容可以看到，3 種交換機的型號分別為 WS-C2960-24PC-L、WS-C2960S-F24PS-L 和 WS-C2960X-24PS-L，可以用正規表示法 switch_model = re.search(r'WS-C2960\w?-\w{4，5}-L' , output) 來比對。

再來看 show flash: | i c2960 的輸出結果，我們獲得了對應 3 種交換機型號的 IOS 的 .bin 檔案名稱分別為 c2960-lanbasek9-mz.122-55.SE12.bin、c2960s-universalk9-mz.150-2. SE11.bin 和 c2960x-universalk9-mz.152-2. E8.bin，可以用正規表示法 ios_version = re.search(r'c2960\ w?-\w{8，10}\d?-mz.\d{3}-\d{1，2}.\w{2，4}(.bin)?'，output) 來比對。

同理，最後的 show boot | i BOOT path 分別獲得了 flash:c2960-lanbasek9-mz.122-55. SE12.bin、flash:c2960s-universalk9-mz.150-2.SE11.bin 和 flash:/c2960x-universalk9-mz.152-2. E8.bin 3 種不同的 boot_system 啟動路徑，用正規表示法 boot_system = re.search(r'flash:/? c2960\w?-\w{9，11}-mz.\d{3}-\d{1，2}.\w{2，4}.bin'，output) 來比對。如果你足夠細心，則會發現 boot_system 路徑可以寫成 flash:，也可以寫成 flash:/，有些網路工程師喜歡加 /，有些網路工程師不喜歡加，所以在正規表示法中我們用 flash:/? 將兩種情況都比對到了。

```
command.send("show inventory | i PID: WS\n")
time.sleep(0.5)
command.send("show flash: | i c2960\n")
time.sleep(0.5)
command.send("show boot | i BOOT path\n")
time.sleep(0.5)
output = command.recv(65535)
command.send("wr mem\n")
switch_model = re.search(r'WS-C2960\w?-\w{4,5}-L'，output)
ios_version = re.search(r'c2960\w?-\w{8,10}\d?-mz.\d{3}-\d{1,2}.\
w{2,4}(.bin)?'，output)
boot_system = re.search(r'flash:.+mz.\d{3}-\d{1,2}.\w{2,4}\.bin'，
output)
```

（4）用正規表示法比對所有可能出現的輸出結果後，接下來就可以用 if 敘述配合 and 和 or 兩個布林邏輯運算來作判斷，先看第一條比對 2960 交換機的 if 敘述。

```
        if switch_model.group() == "WS-C2960-24PC-L" and ios_version.
group() == "c2960-lanbasek9-mz.122-55.SE12.bin" and boot_system.group() ==
'flash:c2960-lanbasek9-mz.122-55.SE12.bin' or boot_system.group() == 'flash:/
c2960-lanbasek9-mz.122-55.SE12.bin':
            switch_upgraded.append(ip_address)
```

如果交換機型號為 WS-2960-24PC-L 且交換機的 IOS 版本為 c2960-
lanbasek9-mz.122- 55.SE12.bin，並且 boot_system 的啟動路徑為 flash:
c2960-lanbasek9-mz.122-55.SE12.bin（不加 /）或 flash:/c2960-lanbasek9-
mz.122-55.SE12.bin（加 /），則將該交換機的管理 IP 位址加入 switch_
upgraded 串列。

同理，再來看第二條比對 2960S 交換機的 elif 敘述。

```
        elif switch_model.group() == "WS-C2960X-24PS-L" and ios_version.
group() == "c2960x-universalk9-mz.152-2.E8.bin" and boot_system.group() ==
'flash:c2960x-universalk9-mz.152-2.E8.bin' or boot_system.group() == 'flash:/
c2960x-universalk9-mz.152-2.E8.bin':
            switch_upgraded.append(ip_address)
```

如果交換機型號為 WS-C2960S-F24PS-L 且交換機的 IOS 版本為 c2960s-
universalk9-mz. 150-2.SE11.bin，並且 flash:c2960s-universalk9-mz.150-2.
SE11.bin（不加 /）或 flash:/c2960s- universalk9-mz.150-2.SE11.bin（加 /），
則將該交換機的管理 IP 位址加入 switch_upgraded 串列。

第三條比對 2960X 交換機的同理，這裡略過不講。

如果任何交換機都不滿足以上 3 條比對條件，則將它的管理 IP 位址加入
switch_not_ upgraded 串列。

```
        elif switch_model.group() == "WS-C2960X-24PS-L" and ios_version.
group() == "c2960x-universalk9-mz.152-2.E8.bin" and boot_system.group() ==
'flash:c2960x-universalk9-mz.152-2.E8.bin' or boot_system.group() == 'flash:/
c2960x-universalk9-mz.152-2.E8.bin':
```

```
        switch_upgraded.append(ip_address)
    else:
        switch_not_upgraded.append(ip_address)
```

（5）最後將 switch_upgraded 和 switch_not_upgraded 兩個串列的所有元素
列印出來，這樣就能清楚地看到哪些交換機已經成功升級，哪些還沒有升
級，以及它們的位址。其餘的程式內容略過不講。

```
    except Paramiko.ssh_exception.AuthenticationException:
        print ("TACACS is not working for " + ip_address + ".")
        switch_with_tacacs_issue.append(ip_address)
    except socket.error:
        print (ip_address +  " is not reachable.")
        switch_not_reachable.append(ip_address)

iplist.close()
ssh_client.close

print ('\nTACACS is not working for below switches: ')
for i in switch_with_tacacs_issue:
    print (i)

print ('\nBelow switches are not reachable: ')
for i in switch_not_reachable:
    print (i)

print ('\nBelow switches IOS version are up-to-date: ')
for i in switch_upgraded:
    print (i)

print ('\nBelow switches IOS version are not updated yet: ')
for i in switch_not_upgraded:
    print (i)
```

5.3.7 指令稿 1 驗證

移動到 /root/lab3，執行指令稿 1（lab3_1.py），指令稿自動登入 ip_list.txt 中所有交換機的管理 IP 位址，然後執行重新啟動交換機前的驗證步驟，可以看到 172.16.206.119 和 172.16.206.146 兩個交換機重新啟動前的 IOS 版本升級設定有誤，如下圖所示。

```
[root@localhost lab3]# python lab3_1.py
Username: parry
Password:
Successfully connect to  172.16.206.39
Successfully connect to  172.16.206.40
Successfully connect to  172.16.206.41
Successfully connect to  172.16.206.42
Successfully connect to  172.16.206.43
Successfully connect to  172.16.206.119
Successfully connect to  172.16.206.128
Successfully connect to  172.16.206.129
Successfully connect to  172.16.206.145
Successfully connect to  172.16.206.146

TACACS is not working for below switches:

Below switches are not reachable:

Below switches IOS version are up-to-date:
172.16.206.39
172.16.206.40
172.16.206.41
172.16.206.42
172.16.206.43
172.16.206.128
172.16.206.129
172.16.206.145

Below switches IOS version are not updated yet:
172.16.206.119
172.16.206.146
```

5.3.8 實驗準備——指令稿 2

我們來看指令稿 2，也就是重新啟動交換機後，驗證其 IOS 版本是否成功升級的指令稿。

首先移動到 lab3 資料夾下，建立 lab3_2.py 指令稿。

```
[root@CentOS-Python ~]# cd lab3
[root@CentOS-Python lab3]# vi lab3_2.py
```

然後在維護視窗時段重新啟動所有交換機，相信讀到這裡的所有讀者都有
能力獨自寫一個指令稿來批次重新啟動交換機了。

5.3.9 實驗程式──指令稿 2

將下列程式寫入指令稿 lab3_2.py。

```python
import paramiko
import time
import getpass
import sys
import re
import socket

username = input("Username: ")
password = getpass.getpass("Password: ")
iplist = open('ip_list.txt','r+')

switch_upgraded = []
switch_not_upgraded = []
switch_with_tacacs_issue = []
switch_not_reachable = []

for line in iplist.readlines():
    try:
        ip_address = line.strip()
        ssh_client = paramiko.SSHClient()
        ssh_client.set_missing_host_key_policy(paramiko.AutoAddPolicy())
        ssh_client.connect(hostname=ip_address,username=username,password=
password)
        print ("Successfully connect to ", ip_address)
        command = ssh_client.invoke_shell(width=300)
```

```
        command.send("show ver | b SW Version\n")
        time.sleep(0.5)
        output = command.recv(65535)
        print (output)
        ios_version = re.search(r'\d{2}.\d\(\d{1,2}\)\w{2,4}', output)
        if ios_version.group() == '12.2(55)SE12':
            switch_upgraded.append(ip_address)
        elif ios_version.group() == '15.2(2)E8':
            switch_upgraded.append(ip_address)
        elif ios_version.group() == '15.0(2)SE11':
            switch_upgraded.append(ip_address)
        else:
            switch_not_upgraded.append(ip_address)
    except paramiko.ssh_exception.AuthenticationException:
        print ("TACACS is not working for " + ip_address + ".")
        switch_with_tacacs_issue.append(ip_address)
    except socket.error:
        print (ip_address +  " is not reachable." )
        switch_not_reachable.append(ip_address)

iplist.close()
ssh_client.close

print ('\nTACACS is not working for below switches: ')
for i in switch_with_tacacs_issue:
    print (i)

print ('\nBelow switches are not reachable: ')
for i in switch_not_reachable:
    print (i)

print ('\nBelow switches IOS version are up-to-date: ')
for i in switch_upgraded:
    print (i)

print ('\nBelow switches IOS version are not updated yet: ')
for i in switch_not_upgraded:
    print (i)
```

5.3.10 指令稿 2 程式分段說明

（1）這段程式和指令稿 1 大致相同，我們同樣建立了 switch_upgraded 和
switch_not_ upgraded 兩個空串列，用來統計有哪些交換機在重新啟動後
IOS 版本升級成功，有哪些交換機在重新啟動後 IOS 版本升級不成功。

```python
import paramiko
import time
import getpass
import sys
import re
import socket

username = input("Username: ")
password = getpass.getpass("Password: ")
iplist = open('ip_list.txt','r+')

switch_upgraded = []
switch_not_upgraded = []
switch_with_tacacs_issue = []
switch_not_reachable = []

for line in iplist.readlines():
    try:
        ip_address = line.strip()
        ssh_client = paramiko.SSHClient()
        ssh_client.set_missing_host_key_policy(paramiko.AutoAddPolicy())
        ssh_client.connect(hostname=ip_address,username=username,password=
password)
        print ("Successfully connect to ", ip_address)
        command = ssh_client.invoke_shell(width=300)
        command.send("show ver | b SW Version\n")
        time.sleep(0.5)
        output = command.recv(65535)
        print (output)
```

（2）重新啟動交換機後直接輸入 show ver | b SW Version 檢視目前的 IOS
版本，再用正規表示法 ios_version = re.search(r'\d{2}.\d\(\d{1，2}\)\w{2，
4}'，output) 來比對，只要能比對到 12.2(55)SE12、15.2(2)E8、15.0(2)SE11
這 3 種 IOS 版本中的任意一種，就將交換機的管理 IP 位址增加到 switch_
upgraded 串列。反之，如果上述 3 種 IOS 版本都比對不到，則將交換機的
管理 IP 位址增加到 switch_not_reachable 串列。最後用 for 循環將 switch_
upgraded 和 switch_not_reachable 中的元素一一列印出來，輸出內容的格式
和指令稿 1 並無二致。

```
        ios_version = re.search(r'\d{2}.\d\(\d{1，2}\)\w{2，4}''，output)
        if ios_version.group() == '12.2(55)SE12':
            switch_upgraded.append(ip_address)
        elif ios_version.group() == '15.2(2)E8':
            switch_upgraded.append(ip_address)
        elif ios_version.group() == '15.0(2)SE11':
            switch_upgraded.append(ip_address)
        else:
            switch_not_upgraded.append(ip_address)
    except Paramiko.ssh_exception.AuthenticationException:
        print ("TACACS is not working for " + ip_address + ".")
        switch_with_tacacs_issue.append(ip_address)
    except socket.error:
        print (ip_address +  " is not reachable." )
        switch_not_reachable.append(ip_address)

iplist.close()
ssh_client.close

print ('\nTACACS is not working for below switches: ')
for i in switch_with_tacacs_issue:
    print (i)

print ('\nBelow switches are not reachable: ')
for i in switch_not_reachable:
```

```
    print (i)

print ('\nBelow switches IOS version are up-to-date: ')
for i in switch_upgraded:
    print (i)

print ('\nBelow switches IOS version are not updated yet: ')
for i in switch_not_upgraded:
    print (i)
```

5.3.11 指令稿 2 驗證

移動至 /root/lab3，執行指令稿 lab3_2.py，可以看到在重新啟動後，所有
交換機的 IOS 版本都已升級成功，包含在執行指令稿 lab3_1.py 時出問題
的 172.16.206.119 和 172.16.206.146 兩台交換機，如下圖所示。

```
[root@localhost lab3]# python lab3_2.py
Username: parry
Password:
Successfully connect to  172.16.206.39
Successfully connect to  172.16.206.40
Successfully connect to  172.16.206.41
Successfully connect to  172.16.206.42
Successfully connect to  172.16.206.43
Successfully connect to  172.16.206.119
Successfully connect to  172.16.206.128
Successfully connect to  172.16.206.129
Successfully connect to  172.16.206.145
Successfully connect to  172.16.206.146

TACACS is not working for below switches:

Below switches are not reachable:

Below switches IOS version are up-to-date:
172.16.206.39
172.16.206.40
172.16.206.41
172.16.206.42
172.16.206.43
172.16.206.119
172.16.206.128
172.16.206.129
172.16.206.145
172.16.206.146

Below switches IOS version are not updated yet:
```

Python 協力廠商模組詳解

在讀完前面 5 章內容並動手操作後,相信讀者已經對 Python 的基礎知識和 Python 在網路運行維護中的應用有了某種程度的了解。在此基礎上,本章將擴充討論 Python 在網路運行維護自動化領域中十分重要的兩個話題:「不支援 API 的傳統裝置」和「Python 預設同步、單執行緒效率不佳」的問題,並將舉例列出解決這些問題的方案及其原理。針對「不支援 API 的傳統裝置」的話題將在 6.1 ～ 6.5 節討論,而在 6.6 和 6.7 兩節將討論如何應對第二個問題。

1. 不支援 API的傳統裝置

眾所皆知,NETCONF(RFC 6241)這個以 XML 為基礎、用來替代 CLI 和 SNMP 的網路設定和網路管理協定最早誕生於 2006 年 12 月,由 IETF(The Internet Engineering Task Force,網際網路工程任務組)以 RFC 4741 發佈。在進行一番修訂後,IETF 又於 2011 年 6 月以 RFC 6241 將其再次發佈。在此之前的一些「上了年紀」的網路裝置,例如思科經典的以 IOS 作業系統為基礎的 Catalyst 系列交換機(2960、3560、3750、4500、6500)是不支援 NETCONF 的,更別提對正在學習 NetDevOps 的網路工程師來說耳熟能詳的 JSON、XML、YAML 這些資料類型，以及 YANG 這個資料模型語言了。簡而言之，這些「古董級」的裝置是沒有 API 的。API 這種「時髦」的東西是思科為了順應 SDN 時代潮流，進而推出新的 IOS-XE 作業系統後逐漸融進思科裝置的，比較有代表性的使用 IOS-XE 的裝置有思科的 3850、9200、9300 等新一代 Catalyst 系列交換機。

由於思科在 21 世紀初期在資料通訊企業佔據「霸主」的地位，加上這些古老的以 IOS 為基礎的裝置確實經典耐用、長盛不衰，因此如今我們仍然能在很多企業和公司的現網裡看到它們的存在，就像今天仍然有一批 Windows XP 和 Windows 7 的忠實使用者在堅守陣地一樣。2016 年 5 月在美國拉斯維加斯舉辦的第 30 屆 Interop 上（Interop 是由英國 Informa 會展公司每年舉辦一次的 IT 技術高峰會），大會嘉賓，Network Programmability and Automation: Skills for the Next-Generation Network Engineer 一書的作者之一，網路運行維護自動化的先驅 Jason Edelman 在他的演講中提到，目前市場上支援 NETCONF、API 的網路裝置只佔市佔率的 15% ～ 20%，剩下的 80% 以上還都是只支援 SSH 存取命令列的傳統裝置。

那麼問題來了：既然 IOS 裝置目前還佔據著相當大的市佔率，那麼是否表示 JSON、XML、YAML 等在這些「古董級」的裝置裡真的毫無用武之地呢？答案是否定的，本章的 6.1 ～ 6.5 節將分別介紹 TextFSM、ntc-template、Napalm、pyntc 這些 Python 的協力廠商模組和範本是如何將 JSON 和這些「老古董們」巧妙結合在一起的，以及它到底能幫我們在這些沒有 API 的裝置上做什麼。在此之前，我們會先介紹一些關於 JSON 的基礎知識。

2. Python 預設同步、單執行緒效率不佳的問題

對網路工程師來說，我們通常必須借助 Telnetlib、Paramiko 或 Netmiko 這些協力廠商開放原始碼模組才能透過 Telnet 或 SSH 來登入、操作和管理各

種網路裝置。因為 Python 預設的執行方式是同步、單執行緒的，也就表示在不對指令稿內容做額外調整的前提下，執行 Python 指令稿的主機只能一台一台地登入裝置執行程式。假設一個指令稿登入一台交換機執行設定平均耗時 5s，那麼在擁有 1000 台交換機的大型企業網裡就要耗時 5000s 才能執行完指令稿，效率太低。在 6.6 和 6.7 節將分別介紹如何透過使用 netdev 和 Netmiko 實現非同步和多執行緒來解決這個痛點問題，提升工作效率。

6.1 JSON

JSON 誕生於 1999 年 12 月，是 JavaScript Programming Language（Standard ECMA-262 3rd Edition）的子集合，是一種輕量級的資料交換格式。雖然 JSON 基於 JavaScript 開發，但它是一種「語言無關」（Language Independent）的文字格式，並且採用 C 語言家族，如 C、C++、C#、Java、Python 和 Perl 等語言的用法習慣，成為一種理想的資料交換語言（Data-Interchange Language）。

6.1.1 JSON 基礎知識

JSON 的資料結構具有易讀的特點，它由鍵值對（Collection）和物件（Object）組成，在結構上非常類似 Python 的字典（Dictionary），以下是一個典型的 JSON 資料格式。

```
{
"intf":"Gigabitethernet0/0",
"status":"up"
}
```

與 Python 的字典一樣，JSON 的鍵值對也由冒號分開，冒號左邊的 "intf" 和 "status" 即鍵值對的鍵（Name），冒號右邊的 "Gigabitethernet0/0" 和 "up" 即鍵值對的值（Value），每組鍵值對之間都用逗點隔開。

JSON 與字典不一樣的地方如下。

（1）**JSON裡鍵的資料類型必須為字串**，而在字典裡字串、常數、浮點數或元組等都能作為鍵的資料類型。

（2）**JSON裡鍵的字串內容必須使用雙引號括起來**，不像字典裡既可以用單引號，又可以用雙引號來表示字串。

JSON 裡鍵值對的值又分為兩種形式，一種形式是簡單的值，包含字串、整數等，例如上面的 "Gigabitethernet0/0" 和 "up" 就是一種簡單的值。另一種形式被稱為物件（Object），物件內容用大括號 {} 表示，物件中的鍵值對之間用逗點分開，**它們是無序的**，舉例如下。

```
{"Vendor":"Cisco", "Model":"2960"}
```

當有多組物件存在時，我們將其稱為 JSON 陣列（JSON Array），陣列以中括號 [] 表示，**陣列中的元素（即各個物件）是有序的（可以把它了解為串列）**，舉例如下。

```
{
    "devices":[
        {"Vendor":"Cisco", "Model":"2960"},
        {"Vendor":"Cisco", "Model":"3560"},
        {"Vendor":"Cisco", "Model":"4500"}
        ]
}
```

6.1.2 JSON 在 Python 中的使用

Python 中已經內建了 JSON 模組，使用時只需 import json 即可，如下圖所示。

```
[root@CentOS-Python ~]# python3.8
Python 3.8.2 (default, Apr 27 2020, 23:06:10)
[GCC 8.3.1 20190507 (Red Hat 8.3.1-4)] on linux
Type "help", "copyright", "credits" or "license" for more information.
>>> import json
>>>
```

JSON 模組主要有兩種函數：json.dumps() 和 json.loads()。前者是 JSON 的編碼器（Encoder），用來將 Python 中的物件轉換成 JSON 格式的字串，如下圖所示。

```
>>> import json
>>>
>>> a = json.dumps('parry')
>>> type(a)
<class 'str'>
>>> type(json.dumps({"c": 0, "b": 0, "a": 0}, sort_keys=True))
<class 'str'>
>>>
>>> type({"c": 0, "b": 0, "a": 0})
<class 'dict'>
>>> type(json.dumps([1,2,3]))
<class 'str'>
```

由此可以看到，我們用 json.dumps() 將 Python 中 3 種類型的物件：字串（'parry'）、字典（{"c": 0, "b": 0, "a": 0}）、串列（[1,2,3]）轉換成了 JSON 格式的字串，並用 type() 函數進行了驗證。

而 json.loads() 的用法則是將 JSON 格式的字串轉換成 Python 的物件，如下圖所示。

```
>>> json_list = '[1,2,3]'
>>> type(json_list)
<class 'str'>
>>> python_list = json.loads(json_list)
>>> print (python_list)
[1, 2, 3]
>>> type(python_list)
<class 'list'>
>>>
>>> json_dictionary = '{"vendor":"Cisco", "model":"2960"}'
>>> python_dictionary = json.loads(json_dictionary)
>>> print (python_dictionary)
{'vendor': 'Cisco', 'model': '2960'}
>>> type(python_dictionary)
<class 'dict'>
>>>
```

我們將兩個 JSON格式的字串："[1,2,3]" 和 "{"vendor":"Cisco", "model":
"2960"}"，用 json.loads() 轉換成了它們各自對應的 Python 物件：串列
[1,2,3] 和字典 {"vendor":"Cisco", "model":"2960"}。需要注意的是，在建立
對應 Python 字典類型的 JSON 字串時，如果對鍵使用了單引號，則 Python
會傳回 "SyntaxError: invalid sytanx"，提示語法錯誤，如下圖所示。

```
>>>
>>> json_dictionary = '{'vendor':'Cisco', 'model':'2960'}'
  File "<stdin>", line 1
    json_dictionary = '{'vendor':'Cisco', 'model':'2960'}'
                          ^
SyntaxError: invalid syntax
>>> exit()
```

原因就是：**JSON 裡鍵的字串內容必須使用雙引號括起來**，不像字典裡既
可以用單引號，又可以用雙引號來表示字串，這點請務必注意。

6.2 正規表示法的痛點

作為本書的重點內容，我們知道，在 Python 中可以使用正規表示法來對
字串格式的文字內容做解析（Parse），進一步比對到我們有興趣的文字內
容，**但是這麼做有一定限制**，例如我們需要用正規表示法從如下圖所示的

一台 2960 交換機 show ip int brief 指令的回應內容中找出哪些**物理通訊埠**
是 Up 的。

```
SW1#show ip int brief
Interface              IP-Address       OK? Method Status         Protocol
GigabitEthernet0/0     unassigned       YES unset  up             up
GigabitEthernet0/1     unassigned       YES unset  down           down
GigabitEthernet0/2     unassigned       YES unset  down           down
GigabitEthernet0/3     unassigned       YES unset  down           down
GigabitEthernet1/0     unassigned       YES unset  down           down
GigabitEthernet1/1     unassigned       YES unset  down           down
GigabitEthernet1/2     unassigned       YES unset  down           down
GigabitEthernet1/3     unassigned       YES unset  down           down
GigabitEthernet2/0     unassigned       YES unset  down           down
GigabitEthernet2/1     unassigned       YES unset  down           down
GigabitEthernet2/2     unassigned       YES unset  down           down
GigabitEthernet2/3     unassigned       YES unset  down           down
GigabitEthernet3/0     unassigned       YES unset  down           down
GigabitEthernet3/1     unassigned       YES unset  down           down
GigabitEthernet3/2     unassigned       YES unset  down           down
GigabitEthernet3/3     unassigned       YES unset  down           down
Vlan1                  192.168.2.11     YES NVRAM  up             up
```

如果用正常思維的正規表示法 GigabitEthernet\d\/\d 來比對，則會將除
VLAN1 外的所有物理通訊埠都比對上，顯然這種做法是錯誤的，因為目前
只有 GigabitEthernet0/0 這一個物理通訊埠的狀態是 Up，如下圖所示。

大家也許此時想到了另一種方法，那就是在交換機的 show ip int brief 後面
加上 | i up 提前做好過濾，再用正規表示法 GigabitEthernet\d\/\d 來比對，
例如下圖這樣。

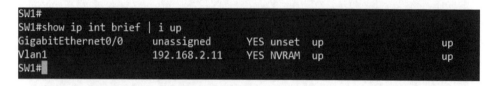

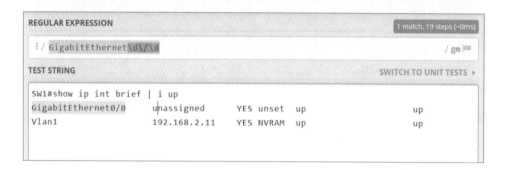

這種方法確實可以比對出我們想要的結果，但是如果這時把要求變一變：
找出目前交換機下所有 Up 的通訊埠（注意，現在不再只是找出所有 Up
的物理通訊埠，虛擬通訊埠也要算進去），並且同時列出它們的通訊埠編
號及 IP 位址。

這個要求表示不僅要用正規表示法比對上面文字裡的 GigabitEthernet0/0、
Vlan1、unassigned、192.168.2.11 這 4 項內容，還要保障正規表示法沒有
比對到其他諸如 YES、unset、NVRAM、up 等文字內容，怎麼樣？這個難
度是不是立即上升了幾個等級？

為了解決類似這樣的正規表示法的痛點問題，我們必須搬出「救兵」—
TextFSM 了。

6.3 TextFSM 和 ntc-templates

TextFSM 最早是由 Google 開發的 Python 開放原始碼模組，它能使用自訂的變數和規則設計出一個範本（Template），然後用該範本來處理文字內容，將這些**無規律的文字內容**按照自己打造的範本將它們整合成想要的**有序的資料格式**。

舉個實例，在 6.2 節的實例中，如果有辦法把 show ip int brief 的回應內容轉換成 JSON 格式，將它們以 JSON 陣列的資料格式列出來，是不是會很方便配合 for 循環比對出我們想要的東西呢？例如看到下面這樣的 JSON 陣列後，讀者是不是想到些什麼呢？

```
[
  {
    "intf": "GigabitEthernet0/0",
    "ipaddr": "unassigned",
    "status": "up",
    "proto": "up"
  },
  {
    "intf": "GigabitEthernet0/1",
    "ipaddr": "unassigned",
    "status": "down",
    "proto": "down"
  },
  {
    "intf": "GigabitEthernet0/2",
    "ipaddr": "unassigned",
    "status": "down",
    "proto": "down"
  },
  {
    "intf": "GigabitEthernet0/3",
```

```
    "ipaddr": "unassigned",
    "status": "down",
    "proto": "down"
  },
]
```

由上面程式可以看到，JSON 陣列格式在 Python 中實際上是一個資料類型為串列的物件（以 [開頭，以] 結尾），既然是串列，那我們就能很方便地使用 for 敘述檢查串列中的每個元素（這裡所有的元素均為字典），並配合 if 敘述，將通訊埠狀態為 "up"（"status" 鍵對應的值）的通訊埠編號（"intf" 鍵對應的值）和 IP 位址（"ipaddr" 鍵對應的值）——列印出來即可（實際的程式將在 6.4 節中列出）。

下面介紹如何在 Python 中使用 TextFSM 建立我們需要的範本。

6.3.1 TextFSM 的安裝

作為 Python 的協力廠商模組，TextFSM 有兩種安裝方法。

第一種方法是使用 pip 安裝，如下圖所示。

```
[root@CentOS-Python templates]# pip3.8 install textfsm
Requirement already satisfied: textfsm in /usr/local/lib/python3.8/site-packages (1.1.0)
Requirement already satisfied: six in /usr/local/lib/python3.8/site-packages (from textfsm) (1.14.0)
Requirement already satisfied: future in /usr/local/lib/python3.8/site-packages (from textfsm) (0.18.2)
WARNING: You are using pip version 19.2.3, however version 20.1 is available.
You should consider upgrading via the 'pip install --upgrade pip' command.
[root@CentOS-Python templates]# _
```

pip 安裝完畢後，進入 Python 並輸入 import textfsm，如果沒有顯示出錯，則說明安裝成功，如下圖所示。

```
[root@CentOS-Python templates]#
[root@CentOS-Python templates]# python3.8
Python 3.8.2 (default, Apr 27 2020, 23:06:10)
[GCC 8.3.1 20190507 (Red Hat 8.3.1-4)] on linux
Type "help", "copyright", "credits" or "license" for more information.
>>> import textfsm
>>> _
```

第二種方法是使用 git clone 指令從 GitHub 下載 TextFSM 的原始程式，如果你的 CentOS 8 主機沒有安裝 Git，系統會提醒你一併安裝，實際如下圖所示。

```
Git clone https://github.com/google/textfsm.git
```

```
[root@CentOS-Python python]# git clone https://github.com/google/textfsm.git
Cloning into 'textfsm'...
remote: Enumerating objects: 150, done.
remote: Counting objects: 100% (150/150), done.
remote: Compressing objects: 100% (112/112), done.
remote: Total 346 (delta 90), reused 84 (delta 38), pack-reused 196
Receiving objects: 100% (346/346), 200.28 KiB | 15.00 KiB/s, done.
Resolving deltas: 100% (192/192), done.
```

原始程式下載後，會看到目前的目錄下多出一個 textfsm 資料夾。進入該資料夾後，會看到如下圖所示的 setup.py 檔案。

```
[root@CentOS-Python python]# ls
ip_list.txt  naplm_test.py  ntc-template.py  __pycache__  textfsm
[root@CentOS-Python python]# cd textfsm
[root@CentOS-Python textfsm]# ls
COPYING  examples  LICENSE.txt  MANIFEST.in  README.md  setup.cfg  setup.py  testdata  tests  textfsm
```

然後輸入下面的指令執行該 py 檔案進行安裝，如下圖所示。

```
python3.8 setup.py install
```

```
[root@CentOS-Python textfsm]# python3.8 setup.py install
running install
running bdist_egg
running egg_info
writing textfsm.egg-info/PKG-INFO
writing dependency_links to textfsm.egg-info/dependency_links.txt
writing entry points to textfsm.egg-info/entry_points.txt
writing requirements to textfsm.egg-info/requires.txt
writing top-level names to textfsm.egg-info/top_level.txt
reading manifest file 'textfsm.egg-info/SOURCES.txt'
reading manifest template 'MANIFEST.in'
writing manifest file 'textfsm.egg-info/SOURCES.txt'
installing library code to build/bdist.linux-x86_64/egg
running install_lib
running build_py
creating build/bdist.linux-x86_64/egg
creating build/bdist.linux-x86_64/egg/textfsm
```

6.3.2 TextFSM 範本的建立和應用

TextFSM 的語法本身並不難，但是使用者必須熟練掌握正規表示法。下面我們以思科 Nexus 7000 交換機上 show vlan 指令的回應內容為例，如下所示，來看如何用 TextFSM 建立範本，以及如何使用範本將該 show vlan 指令的回應內容整理成我們想要的資料格式。

```
N7K# show vlan

VLAN Name                             Status   Ports
---- -------------------------------- -------- ------------------------------
1    default                          active   Eth1/1, Eth1/2, Eth1/3
                                               Eth1/5, Eth1/6, Eth1/7
2    VLAN0002                         active   Po100, Eth1/49, Eth1/50
3    VLAN0003                         active   Po100, Eth1/49, Eth1/50
4    VLAN0004                         active   Po100, Eth1/49, Eth1/50
5    VLAN0005                         active   Po100, Eth1/49, Eth1/50
6    VLAN0006                         active   Po100, Eth1/49, Eth1/50
7    VLAN0007                         active   Po100, Eth1/49, Eth1/50
8    VLAN0008                         active   Po100, Eth1/49, Eth1/50
```

我們建立一個 TextFSM 範本，範本內容如下。

```
Value VLAN_ID (\d+)
Value NAME (\w+)
Value STATUS (\w+)

Start
  ^${VLAN_ID}\s+${NAME}\s+${STATUS}\s+ -> Record
```

（1）在 TextFSM 中，我們使用 Value 敘述來定義變數，這裡定義了 3 個變數，分別是 VLAN_ID、NAME 和 STATUS。

```
Value VLAN_ID (\d+)
Value NAME (\w+)
Value STATUS (\w+)
```

（2）每個變數後面都有它自己對應的正規表示法模式（Pattern），這些模式寫在括號 () 中。例如變數 VLAN_ID 顧名思義是要去比對 VLAN 的 ID，所以它後面的正規表示法模式寫為（\d+）。在 3.6 節中講過，\d 這個特殊序列用來比對數字，後面的 + 用來做貪婪比對。同理，變數 NAME 是用來比對 VLAN 的名稱，因為這裡 VLAN 的名稱摻雜了字母和數字，例如 VLAN0002，所以它的正規表示法模式寫為（\w+）。\w 這個特殊序列用來比對字母或數字，後面的 + 用來做貪婪比對。同理，變數 STATUS（\w+）用來比對 VLAN 狀態，VLAN 狀態會有 active 和 inactive 之分。

（3）在定義好變數後，我們使用 Start 敘述來定義比對規則，比對規則由正規表示法的模式及變數名稱組成。

```
Start
  ^${VLAN_ID}\s+${NAME}\s+${STATUS}\s+ -> Record
```

（4）**Start 敘述後面必須以正規表示法 ^ 開頭。^ 是正規表示法中的一種特殊字元，用於比對輸入字串的開始位置，注意緊隨其後的 $ 不是正規表示法裡的 $，它的作用不是用來比對輸入字串的結尾位置，而是用來呼叫我們之前設定好的 VLAN_ID 並比對該變數。**注意，在 TextFSM 中呼叫變數時可以用大括號 {}，寫成 ${VLAN_ID}，也可以不用，寫成 $VLAN_ID，但是 TextFSM 官方推薦使用大括號。VLAN_ID 對應的正規表示法恰巧是 \d+，這樣就比對到了 1、2、3、4、5、6、7、8 這些 VLAN_ID，而後面的 \s+$ 則表示比對 1、2、3、4、5、6、7、8 後面的空白字元（\s 這個特殊序列用來比對空白字元）。

```
^${VLAN_ID}\s+
```

（5）同理，我們呼叫變數 NAME，它對應的正規表示法模式為 \w+，該特殊序列用來比對 show vlan 指令回應內容中的 default、VLAN0002、

VLAN0003、…、VLAN0008 等內容，而後面的 \s 則用來比對之後所有的空白字元。

```
${NAME}\s+
```

（6）變數 {STATUS} 也一樣，它對應的 \w+ 用來比對 active 和 inactive 這兩種 VLAN 狀態（實例中列出的 show vlan 的回應內容中沒有 inactive，但是不影響了解）以及後面的空白字元。最後用 -> Record 來結束 TextFSM 的比對規則。

```
${STATUS}\s+ -> Record
```

在了解了 TextFSM 的語法基礎後，接下來看怎麼在 Python 中使用 TextFSM。首先將上面的 TextFSM 範本檔案以檔案名稱 show_vlan.template 儲存，如下圖所示。

```
[root@CentOS-Python python]# cat show_vlan.template
Value VLAN_ID (\d+)
Value NAME (\w+)
Value STATUS (\w+)

Start
  ^&{VLAN_ID}\s+${NAME}\s+${STATUS}\s+ -> Record
[root@CentOS-Python python]#
```

然後在相同的資料夾下建立一個名叫 textfsm_demo.py 的 Python 指令稿。

```
[root@CentOS-Python python]#
[root@CentOS-Python python]# vi textfsm_demo.py
```

將下面的程式寫入該指令稿。

```
from textfsm import TextFSM

output = '''
N7K# show vlan

VLAN Name                         Status    Ports
```

```
----  -------------------------------------------------------------
1     default                     active    Eth1/1, Eth1/2, Eth1/3
                                            Eth1/5, Eth1/6, Eth1/7
2     VLAN0002                    active    Po100, Eth1/49, Eth1/50
3     VLAN0003                    active    Po100, Eth1/49, Eth1/50
4     VLAN0004                    active    Po100, Eth1/49, Eth1/50
5     VLAN0005                    active    Po100, Eth1/49, Eth1/50
6     VLAN0006                    active    Po100, Eth1/49, Eth1/50
7     VLAN0007                    active    Po100, Eth1/49, Eth1/50
8     VLAN0008                    active    Po100, Eth1/49, Eth1/50
'''

f = open('show_vlan.template')
template = TextFSM(f)

print (template.ParseText(output))
```

程式分段說明如下。

（1）首 先 我 們 用 from textfsm import TextFSM 引 用 TextFSM 模 組 的
TextFSM 函數，該函數為 TextFSM 模組下最核心的類別。

```
from textfsm import TextFSM
```

（2）將 show vlan 的回應內容以三引號字串的形式設定值給變數 output。

```
output = '''
N7K# show vlan

VLAN Name                      Status    Ports
-------------------------------------------------------------
1     default                  active    Eth1/1, Eth1/2, Eth1/3
                                         Eth1/5, Eth1/6, Eth1/7
2     VLAN0002                 active    Po100, Eth1/49, Eth1/50
3     VLAN0003                 active    Po100, Eth1/49, Eth1/50
4     VLAN0004                 active    Po100, Eth1/49, Eth1/50
```

```
5      VLAN0005                        active     Po100, Eth1/49, Eth1/50
6      VLAN0006                        active     Po100, Eth1/49, Eth1/50
7      VLAN0007                        active     Po100, Eth1/49, Eth1/50
8      VLAN0008                        active     Po100, Eth1/49, Eth1/50
'''
```

（3）開啟之前建立好的範本檔案 show_vlan.template，呼叫 TextFSM() 函數
將它設定值給變數 template，最後呼叫 template 下的 ParseText() 函數對文
字內容進行解析，ParseText() 函數中的參數 output 即 show vlan 指令的回
應內容，最後用 print() 函數將被範本解析後的回應內容列印出來，看看是
什麼樣的內容。

```
f = open('show_vlan.template')
template = TextFSM(f)


print (template.ParseText(output))
```

一切就緒後，執行指令稿看效果，如下圖所示。

```
[root@CentOS-Python python]# python3.8 textfsm_demo.py
[['1', 'default', 'active'], ['2', 'VLAN0002', 'active'], ['3', 'VLAN0003', 'active'], ['4', 'VLAN0004', 'active'], ['5', 'VLAN0
005', 'active'], ['6', 'VLAN0006', 'active'], ['7', 'VLAN0007', 'active'], ['8', 'VLAN0008', 'active']]
```

由此可以看到，之前無序的純字串文字內容被 TextFSM 範本解析後，已經
被有序的巢狀結構串列替代，方便我們配合 for 循環做很多事情。

6.3.3 ntc-templates

用 TextFSM 製作的範本很好用，但是缺點也很明顯：每個 TextFSM 範本
都只能對應一行 show 或 display 指令的回應內容，而目前每家知名廠商的
網路裝置都有上百種 show 或 display 指令，並且每家廠商的回應內容和格
式都完全不同，有些廠商還有多種不同的作業系統，例如思科就有 IOS、
IOS-XE、IOS-XR、NX-OS、ASA、WLC 等多種 OS 版本，這些版本又有
各自特有的 show 指令，難道我們必須自己動手造輪子，每種作業系統的

每行指令都要靠自己手動寫一個對應的範本嗎？不用擔心，已經有前人幫我們造好了輪子，這就是 ntc-templates。

ntc-templates 是由 Network To Code 團隊用 TextFSM 花費了無數心血開發出來的一套範本集，支援 Cisco IOS、Cisco ASA、Cisco NX-OS、Cisco IOS-XR、Arista、Avaya、Brocade、Checkpoint、Fortinet、Dell、Huawei、Palo Alto 等絕大多數主流廠商的裝置，將在這些裝置裡輸入 show 和 display 指令後獲得的各式各樣的回應文字內容整合成 JSON、XML、YAML 等格式。舉例來說，針對思科的 IOS 裝置，ntc-templates 提供了 show ip int brief、show cdp neighbor、show access-list 等常用的 show 指令對應的 TextFSM 範本（還有數百種思科和其他廠商的 TextFSM 範本等待讀者自行去採擷），如下圖所示。

```
-rw-r--r--. 1 root root    481 May  8 10:02 cisco_ios_dir.textfsm
-rw-r--r--. 1 root root   1470 May  8 10:02 cisco_ios_show_access-list.textfsm
-rw-r--r--. 1 root root    953 May  8 10:02 cisco_ios_show_adjacency_detail.textfsm
-rw-r--r--. 1 root root    310 May  8 10:02 cisco_ios_show_aliases.textfsm
-rw-r--r--. 1 root root    180 May  8 10:02 cisco_ios_show_archive.textfsm
-rw-r--r--. 1 root root    403 May  8 10:02 cisco_ios_show_authentication_sessions.textfsm
-rw-r--r--. 1 root root   1096 May  8 10:02 cisco_ios_show_boot.textfsm
-rw-r--r--. 1 root root    264 May  8 10:02 cisco_ios_show_capability_feature_routing.textfsm
-rw-r--r--. 1 root root    937 May  8 10:02 cisco_ios_show_cdp_neighbors_detail.textfsm
-rw-r--r--. 1 root root    546 May  8 10:02 cisco_ios_show_cdp_neighbors.textfsm
-rw-r--r--. 1 root root    317 May  8 10:02 cisco_ios_show_clock.textfsm
-rw-r--r--. 1 root root   2327 May  8 10:02 cisco_ios_show_controller_t1.textfsm
-rw-r--r--. 1 root root    491 May  8 10:02 cisco_ios_show_dmvpn.textfsm
-rw-r--r--. 1 root root   1367 May  8 10:02 cisco_ios_show_dot1x_all.textfsm
-rw-r--r--. 1 root root    413 May  8 10:02 cisco_ios_show_environment_power_all.textfsm
-rw-r--r--. 1 root root   1603 May  8 10:02 cisco_ios_show_environment_temperature.textfsm
-rw-r--r--. 1 root root   1085 May  8 10:02 cisco_ios_show_etherchannel_summary.textfsm
-rw-r--r--. 1 root root    454 May  8 10:02 cisco_ios_show_hosts_summary.textfsm
-rw-r--r--. 1 root root    401 May  8 10:02 cisco_ios_show_interfaces_description.textfsm
-rw-r--r--. 1 root root    484 May  8 10:02 cisco_ios_show_interfaces_status.textfsm
-rw-r--r--. 1 root root   1181 May  8 10:02 cisco_ios_show_interfaces_switchport.textfsm
-rw-r--r--. 1 root root   2082 May  8 10:02 cisco_ios_show_interfaces.textfsm
-rw-r--r--. 1 root root    312 May  8 10:02 cisco_ios_show_interface_transceiver.textfsm
-rw-r--r--. 1 root root    570 May  8 10:02 cisco_ios_show_inventory.textfsm
-rw-r--r--. 1 root root   2000 May  8 10:02 cisco_ios_show_ip_access-lists.textfsm
-rw-r--r--. 1 root root    525 May  8 10:02 cisco_ios_show_ip_arp.textfsm
-rw-r--r--. 1 root root   1272 May  8 10:02 cisco_ios_show_ip_bgp_neighbors.textfsm
-rw-r--r--. 1 root root    502 May  8 10:02 cisco_ios_show_ip_bgp_summary.textfsm
-rw-r--r--. 1 root root   1954 May  8 10:02 cisco_ios_show_ip_bgp.textfsm
-rw-r--r--. 1 root root    711 May  8 10:02 cisco_ios_show_ip_device_tracking_all.textfsm
-rw-r--r--. 1 root root    663 May  8 10:02 cisco_ios_show_ip_eigrp_neighbors.textfsm
-rw-r--r--. 1 root root   2407 May  8 10:02 cisco_ios_show_ip_eigrp_topology.textfsm
-rwxr-xr-x. 1 root root    459 May  8 10:02 cisco_ios_show_ip_flow_toptalkers.textfsm
-rw-r--r--. 1 root root    288 May  8 10:02 cisco_ios_show_ip_interface_brief.textfsm
-rw-r--r--. 1 root root   1778 May  8 10:02 cisco_ios_show_ip_interface.textfsm
```

我們可以透過第 4 章的 Netmiko 模組來呼叫 TextFSM 和 ntc-tempalte。接下來就以實驗的形式示範如何使用 ntc-template。

（1）首先確認主機的 Python 裡安裝了 Netmiko 模組，如下圖所示。

```
[root@CentOS-Python /]# python3.8
Python 3.8.2 (default, Apr 27 2020, 23:06:10)
[GCC 8.3.1 20190507 (Red Hat 8.3.1-4)] on linux
Type "help", "copyright", "credits" or "license" for more information.
>>> import netmiko
>>>
```

（2）在根目錄下建立一個名為 ntc-template 的資料夾，如下圖所示。

```
[root@CentOS-Python /]#
[root@CentOS-Python /]#
[root@CentOS-Python /]# mkdir /ntc-template
```

（3）移動到該資料夾下，如下圖所示。

```
[root@CentOS-Python ~]# cd /ntc-template
[root@CentOS-Python ntc-template]#
```

（4）然後用 git clone 指令下載 ntc-template.git 檔案，如下圖所示。

```
git clone https://github.com/networktocode/ntc-templates.git
```

```
[root@CentOS-Python ntc-template]# git clone https://github.com/networktocode/ntc-templates.git
Cloning into 'ntc-templates'...
remote: Enumerating objects: 48, done.
remote: Counting objects: 100% (48/48), done.
remote: Compressing objects: 100% (45/45), done.
remote: Total 7554 (delta 21), reused 18 (delta 3), pack-reused 7506
Receiving objects: 100% (7554/7554), 1.97 MiB | 6.00 KiB/s, done.
Resolving deltas: 100% (4142/4142), done.
[root@CentOS-Python ntc-template]#
```

（5）安裝完成後，可以用 ls 指令看到在 /ntc-template 下面多出了另一個 ntc-templates 資料夾，再依次輸入指令 cd ntc-templates/templates/ 和 ls，就能看到全部的 ntc-templates 範本集了，如下圖所示。

（6）因為在 Netmiko 裡是用變數 NET_TEXTFSM來呼叫 ntc-templates 範本集的，所以還要用 export 來設定對應的環境變數，將 ntc-templates 範本集的完整路徑 /ntc-template/ ntc-templates/templates 設定值給變數 NET_TEXTFSM，然後用 echo 指令檢驗，如下圖所示。

```
export NET_TEXTFSM='/ntc-template/ntc-templates/templates'
echo $NET_TEXTFSM
```

註：export 指令只在目前生效，下次 CentOS 重新啟動後用 export 指令設定的環境變數就會故障，可以使用 cd ～指令回到目前使用者的家目錄，然後用 vi 編輯家目錄下的 .bashrc 檔案，如下圖所示。

將 export NET_TEXTFSM='/ntc-template/ntc-templates/templates' 寫在 .bashrc 的最下面並儲存，這樣下次 CentOS 重新啟動後的環境變數設定仍然有效，如下圖所示。

```
# .bashrc

# User specific aliases and functions

alias rm='rm -i'
alias cp='cp -i'
alias mv='mv -i'

# Source global definitions
if [ -f /etc/bashrc ]; then
        . /etc/bashrc
fi

export NET_TEXTFSM='/ntc-template/ntc-templates/templates'
```

（7）一切準備就緒後，建立一個 Python 3 的指令稿，指令稿內容如下。

```python
from Netmiko import ConnectHandler
import json

SW1 = {
    'device_type': 'cisco_ios',
    'ip': '192.168.2.11',
    'username': 'python',
    'password': '123',
}

connect = ConnectHandler(**SW1)
print ("Successfully connected to " + SW1['ip'])
interfaces = connect.send_command('show ip int brief', use_textfsm=True)
print (json.dumps(interfaces, indent=2))
```

由上面程式可以看到，我們匯入 Netmiko，以便配合使用 TextFSM 和 ntc-template（匯入 Netmiko 後就不用再匯入 textfsm 模組了）。在 6.1 節中講到，JSON 是 Python 的內建指令稿，可以直接匯入，關於 Netmiko 的其他基礎用法請參考 4.2.2 節，這裡不再贅述。

注意，我們在 Netmiko 的 send_command() 函數中使用了 use_textfsm 參數，並將其設為 True，表示呼叫 TextFSM，而 TextFSM 呼叫的 ntc-template 範本這個步驟我們已經在第 6 步透過改變環境變數做這裡不用操心。最後用 JSON 的 dumps() 函數將解析後的 show ip int brief 的回應文字內容列印出來，注意後面設定縮排的 indent = 2 不能省掉，如果寫成 print (json.dumps(result))，則 Python 會顯示出錯。

（8）執行指令稿看效果，完美地獲得我們預期想要的效果，如下圖所示。

```
[root@CentOS-Python python]# python3.8 test.py
Sucessfully connected to 192.168.2.11
[
  {
    "intf": "GigabitEthernet0/0",
    "ipaddr": "unassigned",
    "status": "up",
    "proto": "up"
  },
  {
    "intf": "GigabitEthernet0/1",
    "ipaddr": "unassigned",
    "status": "down",
    "proto": "down"
  },
  {
    "intf": "GigabitEthernet0/2",
    "ipaddr": "unassigned",
    "status": "down",
    "proto": "down"
  },
  {
    "intf": "GigabitEthernet0/3",
    "ipaddr": "unassigned",
    "status": "down",
    "proto": "down"
  },
```

（9）將之前的 Python 指令稿修改如下。

```
from Netmiko import ConnectHandler
import json
```

```
SW1 = {
    'device_type': 'cisco_ios',
    'ip': '192.168.2.11',
    'username': 'python',
    'password': '123',
}

connect = ConnectHandler(**SW1)
print ("Sucessfully connected to " + SW1['ip'])
interfaces = connect.send_command('show ip int brief', use_textfsm=True)
for interface in interfaces:
    if interface["status"] == 'up':
        print (f'{interface["intf"]} is up!  IP address: {interface["ipaddr"]}')
```

前面講過的程式內容不再重複。需要注意的是，**在 Netmiko 呼叫 TextFSM 後，send_command() 的傳回值不再是字串，而是串列**。所以這裡的變數 interfaces 的資料類型為串列，串列裡的每個元素又如上一步的 JSON 陣列一樣為 JSON 物件，因此可以配合 for 循環和 if 敘述將所有通訊埠狀態為 Up 的通訊埠及對應的 IP 位址都答應出來。

（10）最後再執行一次指令稿，效果如下圖所示。

```
[root@CentOS-Python python]# python3.8 test.py
Sucessfully connected to 192.168.2.11
GigabitEthernet0/0 is up!  IP address: unassigned
Vlan1 is up!  IP address: 192.168.2.11
```

可以看到，在 ntc-templates 的協助下，我們成功地將 show ip int brief 的回應內容轉換成了 JSON 陣列格式，並使用 for 循環和 if 敘述找出了交換機 192.168.2.11 下所有 Up 的通訊埠，以及對應的 IP 位址資訊，完美地解決了正規表示法的痛點問題。

6.4 NAPALM

前面舉例說明了 JSON 的基礎知識及用法，以及如何在 Netmiko 中使用 TextFSM 和 ntc-templates 來對舊型 IOS 裝置的各種 show 和 display 指令的回應內容進行解析，接下來介紹另一個十分強大的協力廠商開放原始碼模組：NAPALM，看看 NAPALM 是如何在對舊型 IOS 裝置做網路運行維護時提供各種便利的。

6.4.1 什麼是 NAPALM

NAPALM 全稱為 Network Automation and Programmability Abstraction Layer with Multivendor support（在英文中，Napalm 是凝固汽油彈的意思，因此 NAPALM 的圖示是一團火焰包圍著一個序列介面，如下圖所示）。顧名思義，NAPALM 是一種為多廠商的網路裝置提供統一 API 的 Python 函數庫，其原始程式可以在 GitHub 上下載。

NAPALM

NAPALM (Network Automation and Programmability Abstraction Layer with Multivendor support) is a Python library that implements a set of functions to interact with different router vendor devices using a unified API.

NAPALM supports several methods to connect to the devices, to manipulate configurations or to retrieve data.

截至 2020 年 5 月，NAPALM 支援 Cisco、Arista、Juniper 3 家主流網路裝置廠商的 5 種作業系統。

- Cisco IOS
- Cisco IOS-XR
- Cisco NX-OS
- Arista EOS
- Juniper JunOS

6.4.2 NAPALM 的優點

NAPALM 簡單好用，首先它依賴 Netmiko，但是又不需要匯入 Netmiko 便可獨立使用（只需要保障執行指令稿的主機安裝了 Netmiko 即可），因此在指令稿裡節省了很多行程式，**讓整個指令稿更加簡潔容易**。

我們知道 TextFSM 可以配合 ntc-templates，將在網路裝置中輸入各種 show 和 display 指令後獲得的無序的純字串類型的回應內容整合成有序的資料結構（串列類型），方便我們使用 for 循環和 if 敘述對這些文字內容做解析，進一步解決 6.2 節提到的問題。而 NAPALM 提供的各種 API 也可以幫助我們在網路裝置上獲得我們有興趣的裝置資訊和參數等內容，**並且 NAPALM 傳回的資料類型也是串列**。目前 NAPALM 提供的 API 如下圖所示，基本覆蓋了網路運行維護中需要經常關注的各種網路資訊和參數。

	EOS	IOS	IOSXR	JUNOS	NXOS	NXOS_SSH
get_arp_table	✓	✓	✗	✗	✗	✓
get_bgp_config	✓	✓	✓	✓	✗	✗
get_bgp_neighbors	✓	✓	✓	✓	✓	✓
get_bgp_neighbors_detail	✓	✓	✓	✓	✗	✗
get_config	✓	✓	✓	✓	✓	✓
get_environment	✓	✓	✓	✓	✓	✓
get_facts	✓	✓	✓	✓	✓	✓
get_firewall_policies	✗	✗	✗	✗	✗	✗
get_interfaces	✓	✓	✓	✓	✓	✓
get_interfaces_counters	✓	✓	✓	✓	✗	✗
get_interfaces_ip	✓	✓	✓	✓	✓	✓
get_ipv6_neighbors_table	✗	✓	✗	✓	✗	✗
get_lldp_neighbors	✓	✓	✓	✓	✓	✓
get_lldp_neighbors_detail	✓	✓	✓	✓	✓	✓
get_mac_address_table	✓	✓	✓	✓	✓	✓
get_network_instances	✓	✓	✗	✓	✓	✗
get_ntp_peers	✗	✓	✓	✓	✓	✓
get_ntp_servers	✓	✓	✓	✓	✓	✓
get_ntp_stats	✓	✓	✓	✓	✓	✗
get_optics	✓	✓	✗	✓	✗	✗
get_probes_config	✗	✓	✓	✓	✗	✗
get_probes_results	✗	✗	✓	✓	✗	✗
get_route_to	✓	✓	✓	✓	✗	✓
get_snmp_information	✓	✓	✓	✓	✓	✓
get_users	✓	✓	✓	✓	✓	✓
is_alive	✓	✓	✓	✓	✓	✓
ping	✓	✓	✗	✓	✓	✓
traceroute	✓	✓	✓	✓	✓	✓

NAPALM 的 API 分為 Getter 類別和 Configuration 類別。上面所列取得裝置參數的 API 即 Getter 類別，而 Configuration 類別則支援對裝置的設定做取代（Config.replace）、合併（Config.merge）、比對（Compare Config）、原子更換（Atomic Changes）、回覆（Rollback）等操作，功能比 TextFSM 和 ntc-template 更強大。

6.4.3 NAPALM 的缺點

前面講到 ntc-template 支援 Cisco IOS、Cisco ASA、Cisco NX-OS、Cisco IOS-XR、Arista、Avaya、Brocade、Checkpoint、Fortinet、Dell、Huawei、Palo Alto 等絕大多數主流廠商的裝置。而相較於 ntc-template，NAPLAM 目前支援的廠商數還較少。

由於 NAPALM 是被其他開發者按照他們自己的習慣和喜好提前造好的輪子，因此 NAPALM 的 API 傳回的資料格式是統一、固定的，很難滿足有個性化需求的使用者，例如在 6.3.3 節的最後，我們用 TextFSM 配合 ntc-template 找出了一個 24 埠的思科 2960 交換機中目前有哪些通訊埠是 Up 的，以及這些通訊埠對應的 IP 位址，如下圖所示。

```
[root@CentOS-Python python]# python3.8 test.py
Sucessfully connected to 192.168.2.11
GigabitEthernet0/0 is up!  IP address: unassigned
Vlan1 is up!  IP address: 192.168.2.11
```

從執行指令稿後列印出的內容可以看到，這裡 GigabitEthernet0/0 是 Up 的，其 IP 位址為 unassigned（二層通訊埠）。另外，Vlan1 通訊埠也是 Up 的，其 IP 位址是 192.168.2.11。但是 NAPALM 的 get_interfaces_ip() 只能列出交換機裡設定了 IP 位址的通訊埠，那些通訊埠狀態為 Up、但是 IP 位址為 unassigned 的二層通訊埠是不會被解析進去的，滿足不了我們的需求。對同一台交換機使用 NAPALM 的 get_interfaces_ip() 後的輸出結果如下圖所示（缺失了 GigabitEthernet0/0）。

```
[root@CentOS-Python python]# python3 naplm_test.py
{
  "Vlan1": {
    "ipv4": {
      "192.168.2.11": {
        "prefix_length": 24
      }
    }
  }
}
[root@CentOS-Python python]# _
```

也許你會説 ntc-templates 裡的各種範本也是提前造好的輪子，傳回的資料
格式也是統一固定的。是的，所以當我們對回應內容的解析有個性化需求
時，推薦使用 TextFSM 來撰寫自己需要的範本。

6.4.4　NAPALM 的安裝

首先確認你的系統中已經安裝了 Netmiko（可以使用指令 pip freeze | grep
Netmiko，也可以進入 Python 解譯器使用 import Netmiko 來確認），然後透
過 pip 下載安裝 NAPALM，如下圖所示。

```
pip install napalm
```

```
[root@CentOS-Python python]# pip3.8 freeze | grep netmiko
netmiko==3.1.0
[root@CentOS-Python python]# pip3.8 install napalm
Requirement already satisfied: napalm in /usr/local/lib/python3.8/site-packages (3.0.0)
Requirement already satisfied: junos-eznc>=2.2.1 in /usr/local/lib/python3.8/site-packages (from napalm) (2.4.1)
Requirement already satisfied: netaddr in /usr/local/lib/python3.8/site-packages (from napalm) (0.7.19)
Requirement already satisfied: pyeapi>=0.8.2 in /usr/local/lib/python3.8/site-packages (from napalm) (0.8.3)
Requirement already satisfied: netmiko>=3.1.0 in /usr/local/lib/python3.8/site-packages (from napalm) (3.1.0)
Requirement already satisfied: textfsm in /usr/local/lib/python3.8/site-packages (from napalm) (1.1.0)
Requirement already satisfied: cffi>=1.11.3 in /usr/local/lib/python3.8/site-packages (from napalm) (1.14.0)
Requirement already satisfied: scp in /usr/local/lib/python3.8/site-packages (from napalm) (0.13.2)
Requirement already satisfied: future in /usr/local/lib/python3.8/site-packages (from napalm) (0.18.2)
Requirement already satisfied: ciscoconfparse in /usr/local/lib/python3.8/site-packages (from napalm) (1.5.4)
Requirement already satisfied: paramiko>=2.6.0 in /usr/local/lib/python3.8/site-packages (from napalm) (2.7.1)
Requirement already satisfied: pyYAML in /usr/local/lib/python3.8/site-packages (from napalm) (5.3.1)
Requirement already satisfied: requests>=2.7.0 in /usr/local/lib/python3.8/site-packages (from napalm) (2.23.0)
Requirement already satisfied: setuptools>=38.4.0 in /usr/local/lib/python3.8/site-packages (from napalm) (41.2.0)
Requirement already satisfied: jinja2 in /usr/local/lib/python3.8/site-packages (from napalm) (2.11.2)
Requirement already satisfied: lxml>=4.3.0 in /usr/local/lib/python3.8/site-packages (from napalm) (4.5.0)
Requirement already satisfied: six in /usr/local/lib/python3.8/site-packages (from junos-eznc>=2.2.1->napalm) (1.14.0)
Requirement already satisfied: ncclient>=0.6.3 in /usr/local/lib/python3.8/site-packages (from junos-eznc>=2.2.1->napalm) (0.6.7)
Requirement already satisfied: pyserial in /usr/local/lib/python3.8/site-packages (from junos-eznc>=2.2.1->napalm) (3.4)
Requirement already satisfied: ntc-templates in /usr/local/lib/python3.8/site-packages (from junos-eznc>=2.2.1->napalm) (1.4.1)
Requirement already satisfied: transitions in /usr/local/lib/python3.8/site-packages (from junos-eznc>=2.2.1->napalm) (0.8.1)
Requirement already satisfied: pyparsing in /usr/local/lib/python3.8/site-packages (from junos-eznc>=2.2.1->napalm) (2.4.7)
Requirement already satisfied: yamlordereddictloader in /usr/local/lib/python3.8/site-packages (from junos-eznc>=2.2.1->napalm) (0.4.0)
Requirement already satisfied: pycparser in /usr/local/lib/python3.8/site-packages (from cffi>=1.11.3->napalm) (2.20)
Requirement already satisfied: passlib in /usr/local/lib/python3.8/site-packages (from ciscoconfparse->napalm) (1.7.2)
Requirement already satisfied: dnspython in /usr/local/lib/python3.8/site-packages (from ciscoconfparse->napalm) (1.16.0)
Requirement already satisfied: colorama in /usr/local/lib/python3.8/site-packages (from ciscoconfparse->napalm) (0.4.3)
Requirement already satisfied: bcrypt>=3.1.3 in /usr/local/lib/python3.8/site-packages (from paramiko>=2.6.0->napalm) (3.1.7)
Requirement already satisfied: pynacl>=1.0.1 in /usr/local/lib/python3.8/site-packages (from paramiko>=2.6.0->napalm) (1.3.0)
Requirement already satisfied: cryptography>=2.5 in /usr/local/lib/python3.8/site-packages (from paramiko>=2.6.0->napalm) (2.9.2)
Requirement already satisfied: urllib3!=1.25.0,!=1.25.1,<1.26,>=1.21.1 in /usr/local/lib/python3.8/site-packages (from requests>=2.7.0->napalm) (1.25.9)
Requirement already satisfied: idna<3,>=2.5 in /usr/local/lib/python3.8/site-packages (from requests>=2.7.0->napalm) (2.9)
Requirement already satisfied: chardet<4,>=3.0.2 in /usr/local/lib/python3.8/site-packages (from requests>=2.7.0->napalm) (3.0.4)
Requirement already satisfied: certifi>=2017.4.17 in /usr/local/lib/python3.8/site-packages (from requests>=2.7.0->napalm) (2020.4.5.1)
Requirement already satisfied: MarkupSafe>=0.23 in /usr/local/lib/python3.8/site-packages (from jinja2->napalm) (1.1.1)
Requirement already satisfied: terminal in /usr/local/lib/python3.8/site-packages (from ntc-templates->junos-eznc>=2.2.1->napalm) (0.4.0)
WARNING: You are using pip version 19.2.3, however version 20.1 is available.
You should consider upgrading via the 'pip install --upgrade pip' command.
[root@CentOS-Python python]#
```

進入 Python，如果輸入 import napalm 沒有顯示出錯，則說明安裝成功，如下圖所示。

```
[root@CentOS-Python python]# python3.8
Python 3.8.2 (default, Apr 27 2020, 23:06:10)
[GCC 8.3.1 20190507 (Red Hat 8.3.1-4)] on linux
Type "help", "copyright", "credits" or "license" for more information.
>>> import napalm
>>> _
```

6.4.5 NAPALM 的應用

我們以實驗的形式舉兩個實例來分別示範如何使用 NAPALM 的 Getter 類別和 Configuration 類別的 API。

首先建立一個名為 napalm1.py 的 Python 指令稿，如下圖所示。

```
[root@CentOS-Python python]# vi napalm1.py_
```

然後在指令稿裡寫入以下程式。

```
from napalm import get_network_driver

driver = get_network_driver('ios')
SW1 = driver('192.168.2.11','python','123')
SW1.open()

output = SW1.get_arp_table()
print (output)
```

由上面程式可以看到，我們用作實驗的交換機是一台 IP 位址為 192.168.2.11、使用者名稱為 python、密碼為 123 的思科 2960 交換機。因為 NAPALM 支援多廠商裝置不同的作業系統，所以首先需要匯入 NAPALM 的 get_network_driver 類別，用來指定該交換機對應的作業系統 IOS 版本。然後呼叫 NAPALM 的 open() 方法，即完成了 SSH 遠端登入交換機的操作。

接著對交換機呼叫 NAPALM 的 get_arp_table() 方法，將它設定值給變數
output，最後將輸出結果列印出來。

執行程式，效果如下圖所示。

```
[root@CentOS-Python python]# python3.8 napalm1.py
[{'interface': 'Vlan1', 'mac': '00:0C:29:9E:A6:8A', 'ip': '192.168.2.1', 'age':
51.0}, {'interface': 'Vlan1', 'mac': '0C:BA:9D:CE:80:01', 'ip': '192.168.2.11',
'age': 0.0}]
[root@CentOS-Python python]# _
```

可以看到，我們**僅使用 6 行程式**就將交換機 show ip arp 指令的回應內容轉
換成了一個有序的串列形式的資料結構，**NAPALM 是不是非常簡單好用**？

如果你覺得輸出結果不易讀，那麼我們也可以配合 JSON 模組在指令稿裡
增加兩行程式，將輸出內容轉換成 JSON 陣列，程式如下。

```
from napalm import get_network_driver
import json

driver = get_network_driver('ios')
SW1 = driver('192.168.2.11','python','123')
SW1.open()

output = SW1.get_arp_table()
print (json.dumps(output, indent=2))
```

再次執行指令稿看效果，如下圖所示。

```
[root@CentOS-Python python]# python3.8 napalm1.py
[
  {
    "interface": "Vlan1",
    "mac": "00:0C:29:9E:A6:8A",
    "ip": "192.168.2.1",
    "age": 54.0
  },
  {
    "interface": "Vlan1",
    "mac": "0C:BA:9D:CE:80:01",
    "ip": "192.168.2.11",
    "age": 0.0
  }
]
[root@CentOS-Python python]# _
```

有關 NAPALM 的 Getter 類別 API 就介紹到這裡，其他比較常用的 Getter 類別 API 包含 get_facts()、get_config()、get_interaces()、get_bgp_config()、get_bgp_interfaces() 等，就留給讀者自行去嘗試和使用吧。

接下來示範如何使用 NAPALM 的 Configuration 類別 API。前面提到，Configuration 類別 API 包含取代（Config.replace）、合併（Config.merge）、比對（Compare Config）、原子更換（Atomic Changes）、回覆（Rollback）5 項操作，這裡舉例說明最常用的合併和比對的用法。

Configuration 類別的合併實際就是給裝置做設定，它的方法是首先建立一個副檔名為 .cfg 的設定檔，將設定指令寫入該設定檔；然後在指令稿裡使用 NAPALM 的 load_merge_candidate() 函數讀取設定檔中的指令並上傳到目標裝置；接著透過 commit_config() 將這些指令在裝置上設定執行。load_merge_candidate() 函數是以 SCP 協定為基礎向目標裝置傳送設定指令的，因此在使用之前，需要先在裝置上開啟 scp server，否則 NAPALM 會傳回 "napalm.base.exceptions.CommandErrorException: SCP file transfers are not enabled. Configure 'ip scp server enable' on the device." 的錯誤，如下圖所示。

```
[root@CentOS-Python python]# python3 napalm2.py
Traceback (most recent call last):
  File "napalm2.py", line 7, in <module>
    iosv12.load_merge_candidate(filename='napalm_config.cfg')
  File "/usr/local/lib/python3.6/site-packages/napalm/ios/ios.py", line 319, in load_merge_candidate
    file_system=self.dest_file_system,
  File "/usr/local/lib/python3.6/site-packages/napalm/ios/ios.py", line 285, in _load_candidate_wrapper
    file_system=file_system,
  File "/usr/local/lib/python3.6/site-packages/napalm/ios/ios.py", line 624, in _scp_file
    TransferClass=FileTransfer,
  File "/usr/local/lib/python3.6/site-packages/napalm/ios/ios.py", line 689, in _xfer_file
    raise CommandErrorException(msg)
napalm.base.exceptions.CommandErrorException: SCP file transfers are not enabled. Configure 'ip scp server enable' on the device.
```

```
SW1#conf t
Enter configuration commands, one per line.  End with CNTL/Z.
SW1(config)#ip scp server enable
```

在了解原理後，首先建立一個名為 napalm_config.cfg 的設定檔，將用來對交換機 VTY 的第 5 ～ 15 行線做基本設定的指令集寫進該設定檔並儲存，如下圖所示。

```
[root@CentOS-Python python]# cat napalm_config.cfg
line vty 5 15
transport input ssh
transport output ssh
login local
```

然後建立一個名為 napalm2.py 的指令稿,如下圖所示。

```
[root@CentOS-Python python]# vi napalm2.py_
```

將下面的程式寫入該指令稿。

```
from napalm import get_network_driver

driver = get_network_driver('ios')
SW1 = driver('192.168.2.11', 'python', '123')
SW1.open()

SW1.load_merge_candidate(filename='napalm_config.cfg')
SW1.commit_config()
```

由上面程式可知,我們用 load_merge_candidate(filename='napalm_config.
cfg') 載入之前建立好的設定檔 napalm_config.cfg,然後配合 NAPALM 的
commit_config() 方法執行該設定檔的設定指令,即完成了對交換機的設定
任務。

在執行程式前,我們先檢查一遍交換機 192.168.2.11 的設定,確認目前交
換機沒有 line vty 5 15 的設定,如下圖所示。

```
SW1#show run | s line vty
line vty 0 4
 login local
 transport input ssh
 transport output ssh
```

然後執行指令稿（因為程式裡沒有使用任何 print() 函數，所以執行程式後沒有任何回應內容），如下圖所示。

```
[root@CentOS-Python python]# python3 napalm2.py
```

接著回到交換機檢查設定，如下圖所示。

```
SW1#show run | s line vty
line vty 0 4
 login local
 transport input ssh
 transport output ssh
line vty 5 15
 login local
 transport input ssh
 transport output ssh
SW1#
SW1#show start | s line vty
line vty 0 4
 login local
 transport input ssh
 transport output ssh
line vty 5 15
 login local
 transport input ssh
 transport output ssh
SW1#
```

可以看到執行指令稿後，NAPALM 已經完成了 VTY 5 ～ 15 線的設定，並且 show start | s line vty 指令的回應內容證實了在完成設定的同時，NAPALM 幫我們儲存了該設定。

下面來看如何使用 Configuration 類別的比對。首先我們手動將 line vty 5 15 下面的 login local、transport input ssh、transport output ssh 3 行指令都從交換機裡移除，移除之後再輸入 show run | s line vty 指令，可以看到之前的 3 行指令已分別被 no login、transport input none 和 transport output none 替代，如下圖所示。

```
SW1(config)#line vty 5 15
SW1(config-line)#no login local
SW1(config-line)#no transport input
SW1(config-line)#no transport output
SW1(config-line)#end
SW1#
*May 15 10:22:05.047: %SYS-5-CONFIG_I: Configured from console by console
SW1#show run | s line vty
line vty 0 4
 login local
 transport input ssh
 transport output ssh
line vty 5 15
 no login
 transport input none
 transport output none
SW1#
```

然後建立一個名為 napalm3.py 的指令稿，如下圖所示。

```
[root@CentOS-Python python]# vi napalm3.py
```

將下面的程式寫入該指令稿。

```python
from napalm import get_network_driver

driver = get_network_driver('ios')
SW1 = driver('192.168.2.11', 'python', '123')
SW1.open()

SW1.load_merge_candidate(filename='napalm_config.cfg')

differences = SW1.compare_config()
if len(differences) > 0:
    print(differences)
    SW1.commit_config()
else:
    print('No changes needed.')
    SW1.discard_config()
```

由程式可知，我們照例用 load_merge_candidate(filename='napalm_config.cfg') 載入之前建立好的設定檔 napalm_config.cfg，然後呼叫 NAPALM 的 compare_config() 方法。顧名思義，compare_config() 是 NAPALM 用來將 napalm_config.cfg 檔案裡的設定和交換機目前的設定做比對用的，它的傳回值（即比對的結果）的資料類型為字串。我們將該比對結果值設定給變數 difference，然後用 if 敘述配合 len() 函數比較對的結果 difference 進行分析，如果 len() 傳回的整數大於 0，則說明 difference 的內容為不可為空，也說明設定檔的指令和交換機目前的設定有區別，我們將比對的結果列印出來，再呼叫 commit_config() 將漏掉的指令補全。反之，如果 len() 傳回的整數為 0，則說明設定檔裡的指令和交換機目前的設定一致，列印 "No changes needed." 提醒使用者不需要對交換機設定做任何更改，並呼叫 NAPALM 的 discard_config() 方法來放棄之前透過 load_merge_candidate(filename='napalm_ config.cfg') 從設定檔 napalm_config.cfg 裡載入好的設定指令。

執行指令稿看效果，如下圖所示。

```
[root@CentOS-Python python]# python3 napalm3.py
+transport input ssh
+transport output ssh
+login local
[root@CentOS-Python python]# _
```

大家注意到 3 行指令前的 3 個加號了嗎？它代表比對的結果變數 difference 的實際內容，表示 compare_config() 方法在將設定檔裡的設定指令和交換機現有的設定做比對後，發現交換機還缺少這 3 行指令，然後透過 SW1.commit_config() 將這 3 行指令在交換機裡補全。我們登入交換機來進行驗證，如下圖所示。

```
SW1#show run | s line vty
line vty 0 4
 login local
 transport input ssh
 transport output ssh
line vty 5 15
 login local
 transport input ssh
 transport output ssh
SW1#
```

可以發現，NAPALM 將交換機缺少的 login local、transport input ssh 和 transport output ssh 這 3 行指令又被補回來了。

6.5 pyntc

除了 NAPALM，pyntc 也是一個非常優秀的用來管理網路裝置設定、升級網路裝置 OS、重新啟動網路裝置的協力廠商開放原始碼模組。也許從名字上你已經猜出來了，是的，同 ntc-templates 一樣，pyntc 也是由 Network to Code 團隊開發製作的，其原始程式儲存在 GitHub 上。

6.5.1 pyntc 和 NAPALM 的比較

同 NAPALM 一樣，pyntc 既依賴 Netmiko（如果物件裝置使用的是思科 IOS 作業系統），也可以在指令稿裡獨立使用（只需要保障執行指令稿的主機安裝了 Netmiko 即可）。這個特性使得 pyntc 擁有同 NAPALM 一樣的一大優點：讓網路運行維護的指令稿程式簡潔、容易，便於維護。

與為不同廠商裝置提供統一 API 的 NAPALM 不一樣，pyntc 是一種多廠商、多 API 的模組。截至 2020 年 5 月，Pyntc 支援包含 Cisco、Arista、Juniper 在內的 3 家主流裝置廠商的 4 種作業系統。

- Cisco IOS platforms - uses SSH (Netmiko)

- Cisco NX-OS - uses pynxos (NX-API)
- Arista EOS - uses pyeapi (eAPI)
- Juniper Junos - uses PyEz (NETCONF)

可以看到，針對不同的作業系統，pyntc 提供的 API 是不一樣的。針對使用思科 IOS 的傳統裝置，因為它們不支援 API，所以 pyntc 借助的是 Netmiko 提供的 SSH 功能來對這些裝置進行存取和管理。針對使用 NX-OS 作業系統的思科 Nexus 系列裝置，pyntc 使用的是 NX-OS 附帶的 NX-API。針對另外兩個廠商：Arista 和 Juniper，pyntc 則分別依賴它們附帶的 eAPI 和 NETCONF 來對裝置進行存取和管理。

6.5.2 pyntc 的安裝

作為 Python 的協力廠商模組，pyntc 有兩種安裝方法。

第一種方法是使用 pip 直接安裝，如下圖所示。

```
[root@CentOS-Python ~]# pip3.8 install pyntc
Collecting pyntc
  Using cached https://files.pythonhosted.org/packages/69/11/5845a8062f379259ef8
748ad1e0cc39a6abd4cdb422c5233b4455601d96a/pyntc-0.0.9-py2.py3-none-any.whl
Requirement already satisfied: requests>=2.7.0 in /usr/local/lib/python3.8/site-
packages (from pyntc) (2.23.0)
Collecting pynxos>=0.0.3 (from pyntc)
  Using cached https://files.pythonhosted.org/packages/f6/37/2f48df4db4fd130d448
c23eb8db5af48ae39363f2c7afa70cb4f4c6cc786/pynxos-0.0.5.tar.gz
Requirement already satisfied: paramiko in /usr/local/lib/python3.8/site-package
s (from pyntc) (2.7.1)
Requirement already satisfied: future in /usr/local/lib/python3.8/site-packages
(from pyntc) (0.18.2)
Requirement already satisfied: netmiko in /usr/local/lib/python3.8/site-packages
 (from pyntc) (3.1.0)
Collecting coverage (from pyntc)
  Downloading https://files.pythonhosted.org/packages/52/38/53a28fcbe77f56f92d45
f9961feca85caebe7424b63a10b07aa8607a4f8d/coverage-5.1-cp38-cp38-manylinux1_x86_6
4.whl (229kB)
     |                              | 235kB 232kB/s
Collecting bigsuds (from pyntc)
  Using cached https://files.pythonhosted.org/packages/0b/42/4884a0856c9c969910a
1585915502ad25163330af75f6247d92cfa3c404e/bigsuds-1.0.6.tar.gz
Collecting mock>=1.3 (from pyntc)
```

pip 安裝完畢後，進入 Python 並輸入 import pyntc，如果 Python 沒有顯示出錯，則説明安裝成功，如下圖所示。

```
[root@CentOS-Python ~]# python3.8
Python 3.8.2 (default, Apr 27 2020, 23:06:10)
[GCC 8.3.1 20190507 (Red Hat 8.3.1-4)] on linux
Type "help", "copyright", "credits" or "license" for more information.
>>> import pyntc
>>> _
```

第二種方法是使用 git clone 指令從 GitHub 下載 TextFSM 的原始程式並安裝，如果你的 CentOS 8 主機沒有安裝 Git，則系統會提醒你一併安裝。

```
git clone https://github.com/networktocode/pyntc.git
cd pyntc
python setup.py install
```

6.5.3 pyntc 的應用

我們以實驗的形式來示範如何使用 pyntc 對使用思科 IOS 作業系統的目標裝置完成以下幾項操作。

- 取得目標裝置的基本資訊。
- 對目標裝置進行設定。
- 取得目標裝置的 running config。
- 對目標裝置的 running config 進行備份。
- 重新啟動目標裝置。

1. 取得目標裝置的基本資訊

首先建立一個名為 pyntc1.py 的指令稿，將下列程式寫入該指令稿。

```
import json
from pyntc import ntc_device as NTC
```

```
SW1 = NTC(host='192.168.2.11', username='python', password='123',
device_ type='cisco_ios_ssh')
SW1.open()

print (json.dumps(SW1.facts, indent=4))
SW1.close()
```

（1）ntc_device 是 pyntc 最重要的類別，pyntc 透過它來 SSH 存取目標 IOS
裝置，因為它的名字有點偏長，我們使用 from pyntc import ntc_device as
NTC 將其命名為 NTC 並呼叫它。

（2）前面講到，同 NAPALM 一樣，pyntc 也依賴 Netmiko，因此存取裝置
的基本設定和方法大致是一樣的。需要注意的是，因為我們要存取的目
標裝置是一台思科 IOS 裝置，所以 NTC() 裡的 device_type 參數我們使用
cisco_ios_ssh。

（3）同 NAPALM 一樣，在存取目標裝置的基本參數設定好後，再呼叫
open() 函數即完成 SSH 遠端登入交換機的操作。

（4）pyntc 中的 facts 方法用來讀取目標裝置的廠商、裝置型號、OS 版
本、序號、主機名稱、uptime、通訊埠串列、VLAN 等基本資訊和設定，
為了使輸出的內容更具有可讀性，我們呼叫 json.dumps() 來將輸出內容轉
換成 JSON 格式。

（5）最後使用 close() 方法退出交換機，關閉 SSH 處理程序。

執行指令稿看效果，如下圖所示。

```
[root@CentOS-Python python]# python3.8 pyntc1.py
{
    "model": "",
    "os_version": "",
    "serial_number": "",
    "hostname": "SW1",
    "vendor": "cisco",
    "uptime": 4260,
    "uptime_string": "00:01:11:00",
    "fqdn": "N/A",
    "interfaces": [
        "GigabitEthernet0/0",
        "GigabitEthernet0/1",
        "GigabitEthernet0/2",
        "GigabitEthernet0/3",
        "GigabitEthernet1/0",
        "GigabitEthernet1/1",
        "GigabitEthernet1/2",
        "GigabitEthernet1/3",
        "GigabitEthernet2/0",
        "GigabitEthernet2/1",
        "GigabitEthernet2/2",
        "GigabitEthernet2/3",
        "GigabitEthernet3/0",
        "GigabitEthernet3/1",
        "GigabitEthernet3/2",
        "GigabitEthernet3/3",
        "Vlan1"
    ],
    "vlans": [],
    "cisco_ios_ssh": {
        "config_register": "0x101"
    }
}
[root@CentOS-Python python]#
[root@CentOS-Python python]#
```

註：因為筆者使用的是 GNS3 模擬器上的虛擬交換機（vios_l2-adventerprisek9-m），所以 pyntc 讀取不到 "model"、"os_version"、"serial_number" 幾項資訊，但是影響不大，如果目標裝置為實機，則不會有這個問題。

2. 對目標裝置進行設定

在 pyntc 中，我們可以使用 config() 和 config_list() 對目標裝置進行設定，兩者的區別是前者一次只能對裝置執行一個指令，而後者顧名思義可將多個指令作為元素放入一個串列中，pyntc 依次呼叫串列中的指令對裝置進行設定。

建立一個名為 pyntc2.py 的指令稿，將下列程式寫入該指令稿。

```
from pyntc import ntc_device as NTC

SW1 = NTC(host='192.168.2.11', username='python', password='123', device_type
='cisco_ios_ssh')
SW1.open()

SW1.config('hostname pyntc_SW1')
SW1.config_list(['router ospf 1', 'network 0.0.0.0 255.255.255.255 area 0'])
SW1.close()
```

由程式可知，我們使用 config('hostame pyntc_SW1') 將交換機的主機名稱由 SW1 改為 pyntc_SW1，用 config_list(['router ospf 1', 'network 0.0.0.0 255.255.255.255 area0']) 在交換機上開啟 OSPF。

執行指令稿前，先手動登入交換機，確定目前的主機名稱為 SW1，並且還沒有設定 OSPF，如下圖所示。

```
SW1#              .
SW1#show run | s router ospf
SW1#
```

執行指令稿後，因為沒有使用 print() 函數，所以不會有任何回應內容，如下圖所示。

```
[root@CentOS-Python python]# python3.8 pyntc2.py
[root@CentOS-Python python]# _
```

回到交換機，如下圖所示，發現其主機名稱已經被 pyntc 設定為 pyntc_SW1，並且 OSPF 也已開啟。

```
pyntc_SW1#show run | s router ospf
router ospf 1
 network 0.0.0.0 255.255.255.255 area 0
pyntc_SW1#
```

3. 取得目標裝置的 running config

透過 pyntc 的 running_config 方法，我們可以很方便地取得裝置的 running config。該方法的傳回值的資料類型為字串，可以用 print() 函數列印出來。這裡建立一個名為 pyntc3.py 的指令稿，將下列程式寫入該指令稿。

```
from pyntc import ntc_device as NTC

SW1 = NTC(host='192.168.2.11', username='python', password='123', device_type
='cisco_ios_ssh')
SW1.open()

run = SW1.running_config
print (run)
SW1.close()
```

程式説明部分略過，我們直接來看指令稿執行後的效果（由於回應內容的長度過長，這裡只截取其中一部分），如下圖所示。

```
[root@CentOS-Python python]# python3.8 pyntc3.py
Building configuration...

Current configuration : 3958 bytes
!
! Last configuration change at 07:53:36 UTC Thu May 21 2020 by python
!
version 15.2
service timestamps debug datetime msec
service timestamps log datetime msec
no service password-encryption
service compress-config
```

```
!
hostname pyntc_SW1
!
boot-start-marker
boot-end-marker
!
!
!
username python privilege 15 password 0 123
no aaa new-model
!
!
!
!
!
!
!
!
!
no ip domain-lookup
ip domain-name python.com
ip cef
no ipv6 cef
!
!
file prompt quiet
!
```

4. 對目標裝置的 running config 進行備份

除了取得裝置的 running config，我們也可以透過 pyntc 對裝置的 running config 做備份。在 4.6 節，我們介紹了如何將思科交換機的設定備份在 TFTP 伺服器裡，過程比較煩瑣，而且指令稿的程式也比較多（共 23 行程式）。這裡筆者將示範如何透過 pyntc 的 backup_running_config() 方法，僅使用 5 行程式就將一台交換機的設定備份在本機主機上。

首先將下列程式寫入名為 pyntc4.py 的指令稿。

```
from pyntc import ntc_device as NTC

SW1 = NTC(host='192.168.2.11', username='python', password='123', device_type
='cisco_ios_ssh')
```

```
SW1.open()

SW1.backup_running_config('SW1_config.cfg')
SW1.close()
```

由程式可知，我們呼叫 pyntc 的 backup_running_config() 對交換機的 running config 做備份，將它儲存在一個名為 SW1_config.cfg 的設定檔裡，該設定檔將在執行指令稿的 CentOS 主機的目前的目錄下自動產生。

執行指令稿前，我們首先在主機上透過指令 ls SW_config.cfg 確認目前的目錄下沒有 SW1_config.cfg 檔案，如下圖所示。

```
[root@CentOS-Python python]# ls SW1_config.cfg
ls: cannot access 'SW1_config.cfg': No such file or directory
[root@CentOS-Python python]# _
```

執行指令稿看效果，如下圖所示。

```
[root@CentOS-Python python]# python3.8 pyntc4.py
[root@CentOS-Python python]# ls SW1_config.cfg
SW1_config.cfg
[root@CentOS-Python python]# cat SW1_config.cfg

Building configuration...

Current configuration : 3958 bytes
!
! Last configuration change at 07:53:36 UTC Thu May 21 2020 by python
!
version 15.2
service timestamps debug datetime msec
service timestamps log datetime msec
no service password-encryption
service compress-config
!
hostname pyntc_SW1
!
boot-start-marker
boot-end-marker
!
!
!
username python privilege 15 password 0 123
no aaa new-model
```

```
!
!
!
!
!
!
!
!
!
no ip domain-lookup
ip domain-name python.com
ip cef
no ipv6 cef
```

執行指令稿後，可以看到目前的目錄多出了 SW1_config.cfg 設定檔，檢視該檔案內容，確認的確是交換機 SW1 的 running config，這樣一台思科交換機設定的備份僅透過 5 行程式就完成了。

5. 重新啟動目標裝置

我們來看如何使用 pyntc 重新啟動目標裝置。將下列程式寫入名為 pyntc5. py 的指令稿。

```python
from pyntc import ntc_device as NTC

SW1 = NTC(host='192.168.2.11', username='python', password='123', device_type
='cisco_ios_ssh')
SW1.open()

SW1.save()
SW1.reboot(confirm=True)
```

由程式可知，在重新啟動裝置前，首先要保障目前的設定已被儲存，這裡使用 pyntc 的 save() 方法，它的功能等於思科的 copy run start 指令，隨後呼叫 pyntc 的 reboot() 方法，並將參數 confirm 設為 True，表示確認要對裝置重新啟動。

執行指令稿看效果，如下圖所示。

```
[root@CentOS-Python python]# python3 pyntc5.py
[root@CentOS-Python python]#
```

```
pyntc_SW1#
-Traceback= 1DDC418z 8DC255z 90582Ez 905550z 90535Dz 9014E5z 90211Bz 9020AFz 909578z 9083F7z 9078D7z 908A48z 8D73E3z 886E
D1z 8BADD3z 8B9F87z - Process "SSH Process", CPU hog, PC 0x0090963B

*May 21 09:39:30.891: %GRUB-5-CONFIG_WRITING: GRUB configuration is being updated on disk. Please wait...
pyntc_SW1#
*May 21 09:39:32.037: %GRUB-5-CONFIG_WRITTEN: GRUB configuration was written to disk successfully.
*May 21 09:39:34.379: %SYS-3-CPUHOG: Task is running for (2000)msecs, more than (2000)msecs (0/0),process = SSH Process.
pyntc_SW1#
*May 21 09:39:39.954: %SYS-5-RELOAD: Reload requested by python on vty0 (192.168.2.1). Reload Reason: Reload command.
pyntc_SW1#
 *May 21 09:39:43.085
Reload requested
```

執行指令稿後，可以看到交換機上跳出了記錄檔提示（這裡我們透過
console 存取交換機），證明指令稿對交換機執行了 copy run start 指令，儲
存了目前的設定，隨後對交換機執行 reload 指令，重新啟動裝置，如下圖
所示。

```
pyntc_SW1#show ver | i uptime
pyntc_SW1 uptime is 3 minutes
pyntc_SW1#show run | s router ospf
router ospf 1
 network 0.0.0.0 255.255.255.255 area 0
pyntc_SW1#
```

最後當交換機重新回到命令列時，發現之前設定的交換機的 hostname 及
OSPF 都被儲存了下來。

6.6 netdev（非同步平行）

netdev 是俄羅斯網路運行維護開發工程師 Sergey Yakovlev 開發的一套用
來實現非同步登入和設定網路裝置的協力廠商開放原始碼模組。在說明
它的用法前，首先需要知道什麼是同步（Synchronous），什麼是非同步
（Asynchronous），以及為什麼使用非同步能夠提升日常網路運行維護的工
作效率。

6.6.1 同步與非同步

所謂同步，可以視為每當系統執行完一段程式或函數後，系統都將一直等待該段程式或函數的傳回值或訊息，直到系統接收傳回值或訊息後才繼續執行下一段程式或函數，**在等待傳回值或訊息期間，程式處於阻塞狀態，系統將不做任何事情。**

本書前面所有有關管理多個裝置的實驗中，我們都是將裝置的 IP 位址預先寫入一個名為 ip_list.txt 的文字檔，然後在指令稿裡使用 open() 函數將其開啟，呼叫 readlines() 函數並配合 for 循環讀取每個裝置的 IP 位址，再透過 Paramiko 或 Netmiko 一台接一台裝置地完成 SSH 登入。像這樣 Python 一次只能登入一台裝置，只有完成一台裝置的設定後才能登入下一台裝置繼續設定的方式就是一種典型的「同步」。

而非同步則恰恰相反，系統在執行完一段程式或函數後，不用阻塞性地等待傳回值或訊息，而是繼續執行下一段程式或函數，**在同一時間段裡執行多個任務（而非傻傻地等著一件事情做完並且結果出來後才去做下一件事情），將多個任務平行，進一步加強程式的執行效率。**如果你讀過數學家華羅庚的《統籌方法》，那麼對其中所舉的實例一定不會感到陌生：同樣是沏茶的步驟，因為燒水需要一段時間，你不用等水煮沸了以後才來洗茶杯、倒茶葉（類似同步），而是在等待燒水的過程中就把茶杯洗好、把茶葉倒好，等水燒開了就能直接泡茶喝了，這裡燒水、洗茶杯、倒茶葉 3 個任務是在同一時間段內平行完成的。這就是一種典型的「非同步」。

同步和非同步有一個相同點：它們都是單執行緒下的概念。關於單執行緒和多執行緒的比較會在 6.7 節中討論。

6.6.2 非同步在 Python 中的應用

自從 Python 在 3.4.x 版本起開始支援非同步後,關於非同步的 Python 語法幾經更改,在 Python 3.4、Python 3.5、Python 3.7 中的實現方式有很大不同,本書後面的實例都將以 Python 3.8.2 為基礎來說明非同步的使用。

要了解非同步在 Python 中的應用,必須知道什麼是**程式碼協同**(Coroutine)、什麼是**任務**(Task)、什麼是**可等待物件**(Awaitable Object)。

我們可以把程式碼協同了解為執行緒的最佳化,看成一種微執行緒。它是一種比執行緒更節省資源、效率更高的系統排程機制。非同步就是以程式碼協同實現為基礎的。在 Python 中,實現程式碼協同的模組主要有 asyncio、gevent 和 tornado,使用較多的是 asyncio。首先來看下面的實例。

```
#coding=utf-8
import asyncio
import time

async def main():
    print('hello')
    await asyncio.sleep(1)
    print('world')

print (f"程式於 {time.strftime('%X')} 開始執行")
asyncio.run(main())
print (f"程式於 {time.strftime('%X')} 執行結束")
```

- 在 Python 中,我們透過在 def 敘述前加上 async 敘述將一個函數定義為程式碼協同函數,在上面的實例中,main() 函數被定義為**程式碼協同函數**。
- 這裡的 await asyncio.sleep(1) 表示**臨時中斷**目前的函數 1s。如果程式中還有其他函數,則繼續執行下一個函數,直到下一個函數執行完畢，再傳回來執行 main() 函數。因為除了一個 main() 函數，就沒有其他函數了，所以在 print('hello') 後，main() 函數休眠了 1s，然後繼續 print('world')。

- 程式碼協同函數不是普通的函數，不能直接用 main() 來呼叫，需要使用 asyncio.run(main()) 才能執行該程式碼協同函數。

- 我們配合 time 模組的 strftime() 函數來記錄程式開始前的時間和程式結束後的時間，可以看到總共耗時確實是 1s。

```
[root@CentOS-Python python]# python3.8 1.py
程式於 06:11:02 開始執行
hello
world
程式於 06:11:03 執行結束
[root@CentOS-Python python]# _
```

需要注意的是，不要把 await asyncio.sleep(1) 和 time.sleep(1) 弄混，後者是在同步中使用的休眠操作，前者是在非同步中使用的，因為只有一個 main() 函數需要執行，所以暫時感受不到這兩者的差別，不用著急，繼續看下面的兩個實例。

```
#coding=utf-8
import asyncio
import time

async def say_after(what, delay):
    print(what)
    await asyncio.sleep(delay)

async def main():
    print (f"程式於 {time.strftime('%X')} 執行結束")
    await say_after('hello',1)
    await say_after('world',2)
    print (f"程式於 {time.strftime('%X')} 執行結束")

asyncio.run(main())
```

- 我們在程式碼協同函數 main() 的基礎上加入了另一個函數 say_after()。同樣地，我們用 async 將它定義為程式碼協同函數。
- 我們在 main() 函數中兩次呼叫 say_after() 函數，因為 say_after() 函數是一個程式碼協同函數，因此在呼叫它時，前面必須加上 await。
- 當 main() 函數第一次呼叫 say_after() 函數時，首先列印出 "hello"，然後休眠 1s；第二次呼叫 say_after() 函數時，列印出 "world"，再休眠 2s。兩次呼叫總共花費 3s 來執行完整個程式，如下圖所示。

```
[root@CentOS-Python python]# python3.8 2.py
程式於08:41:16開始執行
hello
world
程式於 08:41:19 執行結束
[root@CentOS-Python python]# _
```

這時你會説，第一次花費 1s，第二次花費 2s，總共 3s 時間，這沒節省時間啊，兩次呼叫的 say_after() 函數並沒有被**平行**啊，這和同步有什麼區別？別急，繼續往下看。

```
#coding=utf-8
import asyncio
import time

async def say_after(what, delay):
    await asyncio.sleep(delay)
    print(what)

async def main():
task1 = asyncio.create_task(say_after('hello',1))
    task2 = asyncio.create_task(say_after('world',2))
    print (f"程式於 {time.strftime('%X')} 開始執行")
    await task1
    await task2
    print (f"程式於 {time.strftime('%X')} 執行結束")

asyncio.run(main())
```

- 要實現非同步平行，需要將程式碼協同函數包裝成一個**任務**，這裡使用 asyncio 的 create_task() 函數將 say_after() 函數包裝了兩次，並分別設定值給 task1 和 task2 兩個變數。然後使用 await 來呼叫 task1 和 task2 兩個任務。
- 執行指令稿後可以看到，因為 task1 和 task2 是並存執行的，所以程式總共耗時 2s 即告完成（06:57:04—06:57:06），如下圖所示。

```
[root@CentOS-Python python]# python3.8 3.py
程式於 06:57:04 開始執行
hello
world
程式於 06:57:06 執行結束
[root@CentOS-Python python]# _
```

最後來看什麼是**可等待物件**。可等待物件的定義很簡單：如果一個物件可以在 await 敘述中使用，那麼它就是**可等待物件**。可等待物件主要有 3 種類型：程式碼協同、任務和 Future。程式碼協同和任務前面已經講過，Future 不在本書的討論範圍內，有興趣的讀者可以自己參閱其他材料深入學習。

6.6.3 netdev 的安裝和應用

也許你會問，既然 Python 3.4 後加入了 asyncio 內建模組，那麼為什麼我們不能用 Paramiko、Netmiko、NAPALM 和 pyntc 來配合它實現非同步呢？很遺憾的是，**由於非同步在 Python 中引用較晚，上述所有這些模組都不支援前面講的可等待物件，也就無法支援非同步平行。**

2019 年 4 月，受 Netmiko 專案的啟發，俄羅斯網路運行維護開發工程師 Sergey Yakovlev 在 Netmiko 的基礎上開發了一個叫作 netdev 的開放原始碼模組。該模組依賴 Netmiko，並且需要至少 Python 3.5 以上才能執行，最大的特點是支援對網路裝置進行非同步登入和操作。截至 2020 年 5 月，netdev 支援 7 家廠商的 12 種作業系統。

- Cisco IOS

- Cisco IOS XE

- Cisco IOS XR

- Cisco ASA

- Cisco NX-OS

- HP Comware (like V1910 too)

- Fujitsu Blade Switches

- Mikrotik RouterOS

- Arista EOS

- Juniper JunOS

- Aruba AOS 6.X

- Aruba AOS 8.X

- Terminal

下面介紹如何安裝和使用 netdev。netdev 可以透過 pip 直接下載安裝（因為 netdev 依賴 Netmiko，下載 netdev 前請確認你的 Python 主機是否已經安裝了 Netmiko），如下圖所示。

```
[root@CentOS-Python python]# pip3.8 install netdev
Requirement already satisfied: netdev in /usr/local/lib/python3.8/site-packages (0.9.3)
Requirement already satisfied: PyYAML<6.0,>=5.1 in /usr/local/lib/python3.8/site-packages (from netdev
) (5.3.1)
Requirement already satisfied: asyncssh<2.0,>=1.15 in /usr/local/lib/python3.8/site-packages (from net
dev) (1.18.0)
Requirement already satisfied: cryptography>=2.7 in /usr/local/lib/python3.8/site-packages (from async
ssh<2.0,>=1.15->netdev) (2.9.2)
Requirement already satisfied: cffi!=1.11.3,>=1.8 in /usr/local/lib/python3.8/site-packages (from cryp
tography>=2.7->asyncssh<2.0,>=1.15->netdev) (1.14.0)
Requirement already satisfied: six>=1.4.1 in /usr/local/lib/python3.8/site-packages (from cryptography
>=2.7->asyncssh<2.0,>=1.15->netdev) (1.14.0)
Requirement already satisfied: pycparser in /usr/local/lib/python3.8/site-packages (from cffi!=1.11.3,
>=1.8->cryptography>=2.7->asyncssh<2.0,>=1.15->netdev) (2.20)
WARNING: You are using pip version 19.2.3, however version 20.1.1 is available.
You should consider upgrading via the 'pip install --upgrade pip' command.
[root@CentOS-Python python]#
```

下載完成後，進入 Python 編輯器，如果輸入 import netdev 沒有顯示出錯，則說明安裝成功，如下圖所示。

```
[root@CentOS-Python python]# python3.8
Python 3.8.2 (default, Apr 27 2020, 23:06:10)
[GCC 8.3.1 20190507 (Red Hat 8.3.1-4)] on linux
Type "help", "copyright", "credits" or "license" for more information.
>>> import netdev
>>> _
```

在使用 netdev 進行非同步作業之前，先來做一個同步和非同步的
比較試驗。首先用傳統的同步方式，透過 Netmiko 對 5 台交換機
（192.168.2.11 ～ 192.168.2.15）的 line vty 5 15 設定 login local，並統計從
指令稿開始執行到執行完成所花費的時間，然後使用 netdev，透過非同步
方式對同樣的 5 台交換機做同樣的設定，並計時，最後將兩次的耗時相比
較，看看孰優孰劣。

傳統的同步方式的指令稿如下（交換機 192.168.2.11 ～ 192.168.2.15 的 IP
位址被儲存在一個名叫 ip_list.txt 的檔案裡）。

```
from Netmiko import ConnectHandler
import time

f = open('ip_list.txt')
start_time = time.time()

for ips in f.readlines():
    ip = ips.strip()
    SW = {
        'device_type': 'cisco_ios',
        'ip': ip,
        'username': 'python',
        'password': '123',
    }
    connect = ConnectHandler(**SW)
    print ("Sucessfully connected to " + SW['ip'])
    config_commands = ['line vty 5 15','login local','exit']
    output = connect.send_config_set(config_commands)
    print (output)
```

```
print ('Time elapsed: %.2f'%(time.time()-start_time))
```

執行指令稿看效果，如下圖所示。

```
[root@CentOS-Python python]# python3.8 sync_netmiko.py
程式於 07:29:55 開始執行

Sucessfully connected to 192.168.2.11
config term
Enter configuration commands, one per line.  End with CNTL/Z.
pyntc_SW1(config)#line vty 5 15
pyntc_SW1(config-line)#login local
pyntc_SW1(config-line)#exit
pyntc_SW1(config)#end
pyntc_SW1#
Sucessfully connected to 192.168.2.12
config term
Enter configuration commands, one per line.  End with CNTL/Z.
SW2(config)#line vty 5 15
SW2(config-line)#login local
SW2(config-line)#exit
SW2(config)#end
SW2#
Sucessfully connected to 192.168.2.13
config term
Enter configuration commands, one per line.  End with CNTL/Z.
SW3(config)#line vty 5 15
SW3(config-line)#login local
SW3(config-line)#exit
SW3(config)#end
SW3#
Sucessfully connected to 192.168.2.14
config term
Enter configuration commands, one per line.  End with CNTL/Z.
SW4(config)#line vty 5 15
SW4(config-line)#login local
SW4(config-line)#exit
SW4(config)#end
SW4#
Sucessfully connected to 192.168.2.15
config term
Enter configuration commands, one per line.  End with CNTL/Z.
SW5(config)#line vty 5 15
SW5(config-line)#login local
SW5(config-line)#exit
SW5(config)#end
SW5#

程式於 07:30:40 執行結束
[root@CentOS-Python python]# _
```

可以看到，透過同步方式一台一台地登入 5 台交換機完成設定總共耗時
45s（07:29:55 – 07:30:40）。

再來看 netdev 的表現，程式如下。

```python
import asyncio
import netdev
import time

async def task(dev):
    async with netdev.create(**dev) as ios:
        commands = ["line vty 5 15", "login local","exit"]
        out = await ios.send_config_set(commands)
        print(out)

async def run():
    devices = []
    f = open('ip_list.txt')
    for ips in f.readlines():
        ip = ips.strip()
        dev = {  'username' : 'python',
                 'password' : '123',
                 'device_type': 'cisco_ios',
                 'host': ip
            }
        devices.append(dev)
    tasks = [task(dev) for dev in devices]
    await asyncio.wait(tasks)

start_time = time.time()
print (f"程式於 {time.strftime('%X')} 開始執行\n")
asyncio.run(run())
print (f"\n程式於 {time.strftime('%X')} 執行結束")
```

執行程式看效果，如下圖所示。

```
[root@CentOS-Python python]# python3.8 netdev2.py
程式於 07:24:55 開始執行

conf t
Enter configuration commands, one per line.  End with CNTL/Z.
SW4(config)#line vty 5 15
SW4(config-line)#login local
SW4(config-line)#exit
SW4(config)#end
SW4#
conf t
Enter configuration commands, one per line.  End with CNTL/Z.
SW3(config)#line vty 5 15
SW3(config-line)#login local
SW3(config-line)#exit
SW3(config)#end
SW3#
conf t
Enter configuration commands, one per line.  End with CNTL/Z.
SW2(config)#line vty 5 15
SW2(config-line)#login local
SW2(config-line)#exit
SW2(config)#end
SW2#
conf t
Enter configuration commands, one per line.  End with CNTL/Z.
SW5(config)#line vty 5 15
SW5(config-line)#login local
SW5(config-line)#exit
SW5(config)#end
SW5#
conf t
Enter configuration commands, one per line.  End with CNTL/Z.
pyntc_SW1(config)#line vty 5 15
pyntc_SW1(config-line)#login local
pyntc_SW1(config-line)#exit
pyntc_SW1(config)#end
pyntc_SW1#

程式於 07:25:00 執行結束
[root@CentOS-Python python]# _
```

可以看到，透過 netdev 提供的非同步方式，我們**僅耗時 5s（07:24:55 –
07:25:00）便執行完指令稿**，對 5 台交換機完成了同樣的設定，同步和非
同步的工作效率差距之大可見一斑！

另外，除了 netdev，pexpect 也支援非同步平行，有興趣的讀者可以自行參
閱其他資料了解。

6.7 Netmiko（多執行緒）

除了上一節講到的單執行緒非同步平行，我們也可以使用多執行緒（Multithreading）來提升 Python 指令稿的工作效率。第 4 章已經介紹了 Netmiko 的來歷及安裝和使用方法，本節主要介紹如何使用 Netmiko 配合 Python 的內建模組 threading 來實現多執行緒執行 Python 指令稿。

6.7.1 單執行緒與多執行緒

在 6.6.1 節，我們參考了數學家華羅庚的《統籌方法》中的實例來說明單執行緒中同步和非同步的差別。其實我們也可以參考同樣的實例來說明單執行緒和多執行緒的差別。在《統籌方法》中講到的沏茶的這個實例中，如果只有一個人來完成燒水、洗茶杯、倒茶葉 3 項任務，此時只有一個工作力，我們就可以把它看成是單執行緒的。**假設我們能找來 3 個人分別負責燒水、洗茶杯、倒茶葉，並且保障 3 個人同時開工做事，那麼我們就可以把它看成是多執行緒的，每一個工作力都代表一個執行緒。**

在電腦的世界中也是一樣，一個程式可以啟用多個執行緒同時完成多個任務，如果一個執行緒阻塞，其他執行緒並不受影響。現在的 CPU 都是多核心，單執行緒只用到其中的一核心，這其實是對硬體資源的一種浪費（當然不可否認的是，隨著時代的進步，現在的 CPU 已經足夠強大，即使只用單核心也能同時應付多個任務，這也是後來 Python 支援非同步的原因）。如果使用多執行緒來執行 Python 指令稿，則不僅能相當大地加強指令稿的執行速度，加強工作效率，並且還能充分利用主機的硬體資源。下面來看如何在 Python 中使用多執行緒。

6.7.2 多執行緒在 Python 中的應用

Python 3 已經內建了 _thread 和 threading 兩個模組來實現多執行緒。相較於 _thread，threading 提供的方法更多而且更常用，因此我們將舉例說明 threading 模組的用法。首先來看下面這段程式。

```python
import threading
import time

def say_after(what, delay):
    print (what)
    time.sleep(delay)
    print (what)

t = threading.Thread(target = say_after, args = ('hello',3))
t.start()
```

- 我們匯入 Python 內建模組 threading 來實現多執行緒。之後定義一個 say_after(what, delay) 函數，該函數包含 what 和 delay 兩個參數，分別用來表示列印的內容及 time.sleep() 休眠的時間。
- 隨後使用 threading 的 Thread() 函數為 say_after(what, delay) 函數建立一個執行緒並將它設定值給變數 t，注意 Thread() 函數的 target 參數對應的是函數名稱（即 say_after），args 對應的是該 say_after 函數的參數，等於 what = 'hello' 和 delay = 3。
- 最後呼叫 threading 中的 start() 來啟動剛剛建立的執行緒。

執行程式看效果，如下圖所示。

```
[root@CentOS-Python python]# python3.8 threading1.py
hello
hello
[root@CentOS-Python python]# _
```

在列印出第一個 "hello" 後，程式因為 time.sleep(3) 休眠了 3s，之後隨即列印出了第二個 "hello"。因為這時我們只執行了 say_after(what, delay) 這一個函數，並且只執行了一次，因此即使我們啟用了多執行緒，也感受不到它和單執行緒的區別。接下來將程式修改如下。

```
#coding=utf-8
import threading
import time


def say_after(what, delay):
    print (what)
    time.sleep(delay)
    print (what)


t = threading.Thread(target = say_after, args = ('hello',3))

print (f"程式於 {time.strftime('%X')} 開始執行")
t.start()
print (f"程式於 {time.strftime('%X')} 執行結束")
```

■ 這一次呼叫 time.strftime() 來嘗試記錄程式執行前和執行後的時間，看看有什麼「意想不到」的結果。

執行程式看效果，如下圖所示。

```
[root@CentOS-Python python]# python3.8 threading2.py
程式於 02:23:44 開始執行
hello
程式於 02:23:44 執行結束
hello
[root@CentOS-Python python]# _
```

這裡大家一定會問為什麼程式在 02:23:44 開始執行，又在同一時間結束？難道不是該休眠 3s 嗎？為什麼明明 print (f" 程式於 {time.strftime('%X')} 開始執行 ") 和 print (f" 程式於 {time.strftime('%X')} 執行結束 ") 分別寫在

t.start() 的上面和下面，但是不等第二個 "hello" 被列印出來，print (f" 程式於 {time.strftime('%X')} 執行結束 ") 就被執行了？

這是因為除了 threading.Thread() 為 say_after() 函數建立的**使用者執行緒**，print(f" 程式於 {time.strftime('%X')} 開始執行 ") 和 print(f" 程式於 {time.strftime('%X')} 執行結束 ") 兩個 print() 函數也共同佔用了公用的**核心執行緒**。也就是說，該指令稿實際上呼叫了兩個執行緒：一個是使用者執行緒，一個是核心執行緒，也就組成了一個多執行緒的環境。**因為分屬不同的執行緒，say_after() 函數和函數之外的兩個 print 敘述是同時執行的，互不干涉**，所以 print(f" 程式於 {time.strftime('%X')} 執行結束 ") 不會像在單執行緒中那樣等到 t.start() 執行完才被執行，而是在 print (f" 程式於 {time.strftime('%X')} 開始執行 ") 被執行後就馬上跟著執行。這也就解釋了為什麼你會看到原本需要休眠 3s 的指令稿會在 02:23:44 同時開始和結束。

如果想要正確捕捉 say_after(what, delay) 函數開始和結束的時間，需要額外使用 threading 模組的 join() 方法，程式如下。

```
#coding=utf-8
import threading
import time

def say_after(what, delay):
    print (what)
    time.sleep(delay)
    print (what)

t = threading.Thread(target = say_after, args = ('hello',3))

print (f"程式於 {time.strftime('%X')} 開始執行")
t.start()
t.join()
print (f"程式於 {time.strftime('%X')} 執行結束")
```

■ 我們只修改了程式的一處地方，即在 t.start() 下面增加了一個 t.join()。**join() 方法的作用是強制阻塞呼叫它的執行緒，直到該執行緒執行完畢或終止（類似單執行緒同步）。** 因為呼叫 join() 方法的變數 t 正是用 threading.Thread() 為 say_after(what, delay) 函數建立的使用者執行緒，所以使用核心執行緒的 print (f" 程式於 {time.strftime('%X')} 執行結束 ") 必須等待該使用者執行緒執行完畢後才能繼續執行，因此指令稿在即時執行的效果會讓你覺得整體還是以單執行緒同步的方式執行的。

執行程式看效果，如下圖所示。

```
[root@CentOS-Python python]# python3.8 threading3.py
程式於 03:49:19 開始執行
hello
hello
程式於 03:49:22 執行結束
```

可以看到，因為呼叫了 join() 方法，在核心執行緒上執行的 print (f" 程式於 {time.strftime ('%X')} 執行結束 ") 必須等待在使用者執行緒上執行的 say_after(what, delay) 函數執行完畢後，才能繼續執行，因此程式前後執行總共花費 3s，類似單執行緒同步的效果。

最後舉一個實例，來看看如何建立多個使用者執行緒並執行，程式如下。

```python
#coding=utf-8
import threading
import time

def say_after(what, delay):
    print (what)
    time.sleep(delay)
    print (what)

print (f"程式於 {time.strftime('%X')} 開始執行\n")
threads = []
```

```
for i in range(1,6):
    t = threading.Thread(target=say_after, name="執行緒" + str(i), args=
('hello',3))
    print(t.name + '開始執行。')
    t.start()
    threads.append(t)

for i in threads:
    i.join()
print (f"\n程式於 {time.strftime('%X')} 執行結束")
```

■ 我們使用 for 循環配合 range(1,6) 建立了 5 個執行緒，並且將它們以多
   執行緒的形式執行，也就是把 say_after(what, delay) 函數以多執行緒的
   形式執行了 5 次。每個執行緒都作為元素加入 threads 空串列，然後使
   用 for 敘述檢查已經有 5 個執行緒的 threads 串列，對其中的每個執行
   緒都呼叫 join() 方法，確保直到它們都執行結束，才執行核心執行緒的
   print (f' 程式於 {time.strftime('%X')} 執行結束 ")。

執行程式看效果，如下圖所示。

```
[root@CentOS-Python python]# python3.8 threading4.py
程式於 05:13:35 開始執行

執行緒1開始執行。
hello
執行緒2開始執行。
hello
執行緒3開始執行。
hello
執行緒4開始執行。
hello
執行緒5開始執行。
hello
hello
hello
hello
hello.
hello

程式於 05:13:38 執行結束
[root@CentOS-Python python]# _
```

可以看到，我們成功地使用多執行緒將程式執行，如果以單執行緒來執行 5 次 say_after(what,delay)函數，則需要花費 3×5=15s 才能執行完整個指令稿，而在多執行緒形式下，整個程式只花費 3s 就執行完畢。

6.7.3 多執行緒在 Netmiko 中的應用

在掌握了 threading 模組的基本用法後，我們來看如何將它與 Netmiko 結合，實現透過 Netmiko 對網路進行多執行緒登入和操作。

在 6.6.3 節中，我們使用傳統的單執行緒同步方式，透過 Netmiko 對 5 台交換機（192.168.2.11 ～ 192.168.2.15）的 line vty 5 15 設定了 login local，整個指令稿從開始執行到結束，前後總共耗時 45.02s。這裡我們用 Netmiko 以多執行緒的方式，對這 5 台交換機做同樣的設定並計時，來看下能節省多少時間。

指令稿如下。

```
#coding=utf-8
import threading
from queue import Queue
import time
from Netmiko import ConnectHandler

f = open('ip_list.txt')
threads = []

def ssh_session(ip, output_q):
    commands = ["line vty 5 15", "login local","exit"]
    SW = {'device_type': 'cisco_ios', 'ip': ip, 'username': 'python',
'password': '123'}
    ssh_session = ConnectHandler(**SW)
    output = ssh_session.send_config_set(commands)
    print (output)
```

```
print (f"程式於 {time.strftime('%X')} 開始執行\n")
for ips in f.readlines():
    t = threading.Thread(target=ssh_session, args=(ips.strip(), Queue()))
    t.start()
    threads.append(t)

for i in threads:
    i.join()
print (f"\n程式於 {time.strftime('%X')} 執行結束")
```

■ 在使用 Netmiko 實現多執行緒時，需要匯入內建佇列模組 queue。在
  Python 中，佇列（queue）是執行緒間最常用的交換資料的形式，這
  裡參考 queue 模組中的 Queue 類別，也就是先進先出（First Input First
  Output，FIFO）佇列，並將它作為出佇列參數設定給 ssh_session(ip,
  output_q) 函數。有關 queue 模組的實際介紹不在本書的討論範圍內，只
  需要知道這是使用 Netmiko 實現多執行緒時必備的步驟即可。

■ 其餘程式的說明從略。

執行指令稿看效果，如下圖所示。

```
[root@CentOS-Python python]# python3.8 threading6.py
程式於 08:11:41 開始執行

config term
Enter configuration commands, one per line.  End with CNTL/Z.
SW5(config)#line vty 5 15
SW5(config-line)#login local
SW5(config-line)#exit
SW5(config)#end
SW5#
config term
Enter configuration commands, one per line.  End with CNTL/Z.
SW4(config)#line vty 5 15
SW4(config-line)#login local
SW4(config-line)#exit
SW4(config)#end
SW4#
```

```
config term
Enter configuration commands, one per line.  End with CNTL/Z.
SW3(config)#line vty 5 15
SW3(config-line)#login local
SW3(config-line)#exit
SW3(config)#end
SW3#
config term
Enter configuration commands, one per line.  End with CNTL/Z.
pyntc_SW1(config)#line vty 5 15
pyntc_SW1(config-line)#login local
pyntc_SW1(config-line)#exit
pyntc_SW1(config)#end
pyntc_SW1#
config term
Enter configuration commands, one per line.  End with CNTL/Z.
SW2(config)#line vty 5 15
SW2(config-line)#login local
SW2(config-line)#exit
SW2(config)#end
SW2#

程式於 08:11:51 執行結束
[root@CentOS-Python python]# _
```

可以看到，使用 Netmiko 多執行緒總共僅耗時 10s（08:11:41 – 08:11:51）
便完成了對 5 台交換機的設定。比預設使用單執行緒同步的傳統方式快了
35s，不過這個速度依然慢於 6.6 節提到的單執行緒非同步方式。在該節實
驗中，我們使用 netdev 單執行緒總共僅耗時 5s 便對同樣的 5 台交換機完
成了同樣的設定。